Lecture Notes in Physics

Volume 897

The Lecture Notes in Physics

The series Lecture Notes in Physics (LNP), founded in 1969, reports new developments in physics research and teaching-quickly and informally, but with a high quality and the explicit aim to summarize and communicate current knowledge in an accessible way. Books published in this series are conceived as bridging material between advanced graduate textbooks and the forefront of research and to serve three purposes:

- to be a compact and modern up-to-date source of reference on a well-defined topic
- to serve as an accessible introduction to the field to postgraduate students and nonspecialist researchers from related areas
- to be a source of advanced teaching material for specialized seminars, courses and schools

Both monographs and multi-author volumes will be considered for publication. Edited volumes should, however, consist of a very limited number of contributions only. Proceedings will not be considered for LNP.

Volumes published in LNP are disseminated both in print and in electronic formats, the electronic archive being available at springerlink.com. The series content is indexed, abstracted and referenced by many abstracting and information services, bibliographic networks, subscription agencies, library networks, and consortia.

Proposals should be sent to a member of the Editorial Board, or directly to the managing editor at Springer:

Christian Caron
Springer Heidelberg
Physics Editorial Department I
Tiergartenstrasse 17
69121 Heidelberg/Germany
christian.caron@springer.com

More information about this series at
http://www.springer.com/series/5304

Herbert Pfister • Markus King

Inertia and Gravitation

The Fundamental Nature and Structure of Space-Time

 Springer

Herbert Pfister
Institut für Theoretische Physik
Universität Tübingen
Tübingen
Germany

Markus King
Fakultät Engineering
Hochschule Albstadt-Sigmaringen
Albstadt
Germany

ISSN 0075-8450   ISSN 1616-6361 (electronic)
Lecture Notes in Physics
ISBN 978-3-319-15035-2   ISBN 978-3-319-15036-9   (eBook)
DOI 10.1007/978-3-319-15036-9

Library of Congress Control Number: 2015933272

Springer Cham Heidelberg New York Dordrecht London

Printed on acid-free paper

Springer International Publishing AG Switzerland is part of Springer Science+Business Media
(www.springer.com)

Preface

This is a book on foundational issues of classical (non-quantum) physics. It is not a systematic textbook, and it contains no exercises. Rather, we try to impart some knowledge about the fundamental nature and structure of the physical spacetime in a mostly non-technical, nevertheless hopefully precise and consistent language. (A few more technical details are deferred to two appendices.) Some basic knowledge of general relativity and differential geometry would certainly be helpful, especially for Chap. 2. In some places we also follow the historical evolution of views and ideas, and we enrich the presentation by characteristic, sometimes provocative quotations from the creators of the associated theories and models. An extensive reference list documents the fact that all the claims and theorems formulated here have a secure mathematical and/or observational basis, and it provides the interested and educated reader with the possibility to dig deeper into special topics. In particular, the book aims to provide a (surely personal) selection of relevant literature on the foundations and the nature of spacetime, and one that is not usually found in standard textbooks on classical mechanics or general relativity. Although the book would not properly fall into the category of popular books, undergraduate students should be able to read at least parts of the book (in particular Chap. 1) with understanding and profit. On the other hand, we hope that even professional experts will find some new viewpoints and connections between different topics. The selection of topics treated in detail (and the omission of other items) may be somewhat unusual, but we think it is justified in view of the title of the book. Wherever possible, we emphasize the mutual dependence and interplay between different subjects.

As the title of the book promises, a central theme is the phenomenon of inertia, from its historic introduction by Galileo and Newton to the Machian hypothesis of its origin in cosmology, and the (at least partial) observational confirmation of this astonishing fact. It is quite obvious that the first part of the book title is adopted from the book 'Gravitation and Inertia' (Ciufolini and Wheeler 1995), but interchanging the words because, presumably more than in any other book, inertia is a dominating theme of the present book. The second part of the title is taken from the brilliant talk by J. Ehlers at the Trieste conference celebrating Dirac's seventieth birthday (Ehlers

1973). In this way, we would also like to honor the two scientists, J.A. Wheeler and J. Ehlers, who have been the most important mentors for one of the present authors (H.P.) in his struggle with and enthusiasm for general relativity.

According to Einstein's equivalence principle, inertia is intimately connected with gravitation, and on the basis of this principle, through a tedious process, Einstein eventually developed his relativistic theory of gravity, general relativity. This theory turned out to be a complete triumph in every respect, and it stands today unchanged, and without a single conflict with experiment, nearly 100 years after its creation. Chapters 2–4 deal with this theory but, as already stressed above, not in the sense of a systematic and complete, textbook-like presentation, but by focusing on some characteristic and fundamental topics within this immensely rich theory. In Chap. 2, we review the remarkable 'derivation' of the (pseudo-)Riemannian spacetime structure from the properties of the elementary objects 'free particles' and 'light rays', as initiated by H. Weyl, and worked out in detail by J. Ehlers, F. Pirani, A. Schild, and others, nowadays known as EPS axiomatics. Such a physically satisfying deduction of the different levels of geometric structure of our 'world' is hardly ever found, even in modern textbook presentations of Einstein's theory of gravity.

Chapter 3 tries to give prominence to the very special structure of general relativity, which manifests itself in the many different routes leading to Einstein's field equations, in deep mathematical results (Cauchy problem, positive energy theorem, singularity theorems), and in spectacular astrophysical predictions, e.g., the existence of black holes. General relativity also encompasses the whole of classical physics, and therefore Einstein's field equations are the most involved, but also the richest equations one can think of in this regime. We hope to transmit some of our own astonishment and fascination for the beauty, consistency, and completeness of this theory, something which nobody could foresee at the time of its creation.

Chapter 4 returns in a way to the central topic of this book, inertia. In their attempt to find the basic source of this phenomenon, E. Mach, B. and I. Friedlaender, A. Föppl, A. Einstein, and others mainly considered rotating systems, in the tradition of Newton's rotating bucket. Within general relativity, this led to different models for the so-called dragging of free particles and inertial systems by accelerating, and particularly rotating heavy masses. Taken as a whole, such examples strengthen the hypothesis of E. Mach that inertia is, in a non-causal way, ruled by the overall masses of the universe. Modern precision experiments and observations are in accordance with this view, for which interesting and stimulating, sometimes provocative formulations can be found in modern textbooks and research articles. A particular effect of moving masses in general relativity shows up in a new, non-Newtonian 'force', called gravitomagnetism, which has recently been confirmed by intricate satellite experiments, 90 years after its prediction by A. Einstein and H. Thirring.

A decisive seed for parts of this book, particularly Chap. 1, and for parts of the research of one of the authors (H.P.), grew out of the following incident. In 1972, as a young lecturer at the University of Tübingen, Germany, he had the duty (or rather the privilege) to deliver a major course on theoretical mechanics (4 h a week, over a whole year). In preparing this course, he studied the standard textbooks on mechanics and was thoroughly dissatisfied with the way the foundations of mechanics, and in particular Newton's first law (the law of inertia), were presented in most of these textbooks. A first shortcoming results from the fact how briefly, superficially, and carelessly this law is often treated. Since this is usually the first law which is presented to young physics students—and it is after all one of the most important and universal laws, with relevance in all areas of physics, not only for mechanics—one would expect, also for pedagogical purposes, that it would serve as a model for a thorough, clear, and logically convincing presentation of the fundamental facts of nature. The difficulty, but also the general failure to achieve this, was clearly expressed long ago by H. Hertz in his famous book on mechanics (Hertz 1894):

> It is quite difficult to present the introduction to mechanics to an intelligent audience without some embarrassment, without the feeling that one should apologize here and there, without the wish to pass quickly over the beginnings.

Besides this fast and careless way of passing over the beginnings, an even more serious deficit in many textbooks shows up in tautologies, circular arguments, a missing distinction between definitions and non-trivial facts of nature, and in some cases in a mixing between Newton's first, second, and third laws. To make this concrete, we quote from a well-established and highly recommended mechanics textbook (Marion 1965, p. 58):

> Thus, the first and second laws are not really 'laws' in the usual sense of the term as used in physics; rather, they may be considered as *definitions*. The third law, on the other hand, is indeed a *law*. It is a statement concerning the real physical world and contains all of the physics in Newton's laws of motion. [Then in a footnote] The reasoning presented here, viz., that the first and second laws are actually definitions and that the third law contains the physics, is not the only possible interpretation. Lindsay and Margenau (1936) for example, present the first two laws as physical laws and then derive the third law as a consequence.

And this unspeakable formulation, which is really an insult to Newton, is to be found, not only in the first edition of the book from 1965, but literally unchanged in the fourth edition from 1995. [We admit that there are a few textbooks that really take care over the presentation of Newton's basic laws, and avoid most of the pitfalls. As an example, we call attention to the textbook Straumann (1987).]

Surely one cannot blame Newton for all the nonsense about Newton's laws in modern textbooks, even if from today's perspective and with today's knowledge some of Newton's concepts and formulations are unfortunate or even untenable, e.g., the concepts of 'absolute space' and 'absolute time'. However, in order to appreciate Newton's genius and his primacy in the formulation of the foundations of physics, one has to compare Newton's *Principia* with the (even less tenable, and much less successful) attempts of his forerunners and contemporaries, as we shall do to some extent in Sect. 1.1. And Newton was well aware of the difficulty of the subject and of the provisional nature of his attempt, when he wrote in the preface to the *Principia* (Newton 1687, p. 383):

I earnestly ask that everything be read with an open mind and that the defects in a subject so difficult may be not so much reprehended as investigated, and kindly supplemented by new endeavors of my readers.

But the unsurpassed success of Newton's program in all applications prevented later generations, and many of today's textbook authors, from examining Newton's formulations critically and from eliminating its defects. For nearly 200 years, there was no real critical reflection and therefore no improvement on Newton's foundations of mechanics, until finally, beginning in the year 1870, a growing number of researchers revisited these questions and reached, besides a fundamental critique of Newton's concepts of 'absolute space' and 'absolute time' (particularly by E. Mach), a genuine clarification, and an elimination of Newton's absolute concepts, mainly due to C. Neumann and L. Lange, as we shall analyze in detail in Sect. 1.2. Sections 1.3 and 1.4 attempt to improve on some deficiencies still present in the work of L. Lange, particularly in the definition of the basic ingredients of the law of inertia, 'free particles', and 'straight lines'.

The geometrical approach to Einstein's relativistic spacetime theory, emphasizing the intrinsic spacetime structure, starting with Weyl and culminating in the EPS axiomatics, is essentially based on a four-dimensional formulation of spacetime as a differential manifold with geometrical structures (defined by tensor fields), the second central theme of this book. In this respect, the book should also serve as an introduction to the conceptual foundations of spacetime theories, and their peculiar character based on 'events' and the spatio-temporal relations between them, and this from the perspective of Newton's law of inertia. Section 1.5 develops such a four-dimensional analysis of Newtonian physics, thereby bringing out the fundamental nature of non-relativistic spacetime (namely, the so-called Leibnizian, Galilean, and Newtonian accounts of space and time). In this respect, this section serves as a preparation for, and contrast to, the corresponding structures in general relativity, presented in Chap. 2. The spacetime geometry of standard Newtonian physics, consistent with Lange's analysis of the law of inertia, is Galilean spacetime, which is "presupposed in all standard accounts of Newtonian mechanics, even though this presupposition is usually tacit and unremarked" (Maudlin 2012, p. 64). Newton's law of gravitation, its reformulation in a four-dimensional geometric setting (in the spirit of special and general relativity), its distinctive properties, and some typical applications in astrophysics, e.g., rotating stars, are treated in Sect. 1.6. And here, surprisingly, even in this centuries old classical field, lie dormant astrophysically relevant, but mathematically difficult and hitherto unsolved problems.

We are deeply indebted to our copy editor Stephen N. Lyle for the careful edit of our manuscript, for the valuable comments, and for his judicious corrections and improvement of grammar and style. We wish also to thank our editor Christian Caron at Springer for excellent cooperation and support of this book project within the Lecture Notes in Physics series.

Tübingen, Germany   H. Pfister
November 2014   M. King

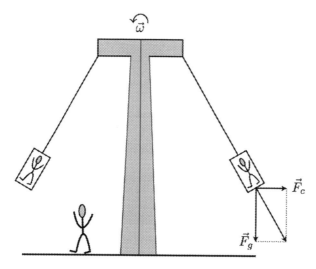

This sketch of a merry-go-round shall serve as an example for the combined action of inertia (centrifugal force $\mathbf{F}_c$) and gravity (force $\mathbf{F}_g$) from everyday life.

Notation

In most cases we use the so-called geometric units, where the gravitational constant G and the light velocity c are set to 1. Four-dimensional spacetime coordinates are denoted by Greek letters $\mu, \nu, \ldots = 0, 1, 2, 3$. Three-dimensional space coordinates are denoted by Latin letters $i, k, \ldots = 1, 2, 3$. The Minkowski metric is denoted by $\eta_{\mu\nu} = \mathrm{diag}(1, -1, -1, -1)$. The symbol $A_{[\mu\nu]}$ means antisymmetrization between the indices, i.e., $A_{[\mu\nu]} = (A_{\mu\nu} - A_{\nu\mu})/2$. Partial differentiation is written $(\ \ )_{,\mu}$, and covariant differentiation is written $(\ \ )_{;\mu}$. Occasionally X is written instead of X^μ for a 4-vector, and $\mathbf{x}$ denotes a 3-vector.

References

Ciufolini, I., Wheeler, J.A.: Gravitation and Inertia. Princeton University Press, Princeton (1995)

Ehlers, J.: The nature and structure of spacetime. In: Mehra J. (ed.) The Physicist's Conception of Nature, pp. 71–91. Reidel, Dordrecht (1973)

Hertz, H.: Die Prinzipien der Mechanik, p. 8. J. A. Barth, Leipzig (1894)

Lindsay, R.B., Margenau, H.: Foundations of Physics. Wiley, New York (1936)

Marion, J.B.: Classical Dynamics of Particles and Systems. Academic, New York (1965); Fourth edition by J.B. Marion and S.T. Thornton, Harcourt Brace, Fort Worth, p. 50 (1995)

Maudlin, T.: Philosophy of Physics: Space and Time. Princeton University Press, Princeton (2012)

Newton, I.: Mathematical Principles of Natural Philosophy (1687). Translated and edited by I.B. Cohen, A. Whitman. University of California Press, Berkeley (1999)

Straumann, N.: Klassische Mechanik. Lecture Notes in Physics, vol. 289. Springer, Berlin (1987); Second enlarged edition published as Theoretische Mechanik. Springer, Berlin (2015)

Contents

Chapter 1
The Laws of Inertia and Gravitation in Newtonian Physics

1.1 Historical Remarks (Galileo, Huygens, Newton)

It seems appropriate to begin the historical analysis of the laws of inertia and gravitation with the work of Galileo. Of course, in the classical Greek period, in the Hellenistic period, and in the early middle ages, there was already much activity and also progress in the natural sciences and in 'physics', a term introduced by Aristotle in around 335 BC. However, in those days, arguments were mostly of a philosophical and qualitative character, and not corroborated by experiment. Moreover, it was more or less a dogma that the phenomena and laws of the cosmos were independent of, and indeed generally different from, those prevailing on Earth. If dynamical processes were analyzed at all, the aim was to find a cause (a 'force') for the velocity of an object, and not for its acceleration, as is the case in modern physics. In short, one can say that the Aristotelian philosophy, so dominant for many centuries, was more of a hindrance than a help in the search for the correct laws of inertia and gravitation. (For a detailed presentation of the classical and Hellenistic period, and of the middle ages, especially from the point of view of absolute versus relative concepts, see, e.g., Barbour 1989, Chaps. 2–4.)

It is generally agreed that Galileo (1564–1642) is the father of the modern way of doing physics: starting from precise observations and from specific experiments— often in an idealized manner by minimizing disturbing effects like friction—one attempts to generate mathematical relations between measurement data, in the ideal case discovering a physical law, and testing such a law by new (different or refined) experiments. Of the many successes Galileo achieved using this method, the following are particularly relevant for our analysis of the laws of inertia and gravitation. By experimenting with pendulums of different material, Galileo found that all bodies fall equally (disregarding friction), or in modern language, that inertial and gravitational mass always coincide. To a large extent from his experiments with balls rolling down an inclined plane, Galileo derived the law of free fall: the free fall height grows quadratically with time. The law of inertia

© Springer International Publishing Switzerland 2015
H. Pfister, M. King, *Inertia and Gravitation*, Lecture Notes in Physics 897,
DOI 10.1007/978-3-319-15036-9_1

(free particles move uniformly on straight lines relative to an inertial system) could not have been formulated by Galileo with this clarity and completeness, mainly because the concept of an inertial system was not yet available at the time. But it was completely clear to him that a body free of any external influence ('force') maintains its constant horizontal velocity. And he combined this constant free horizontal motion with his law of free fall, applying the parallelogram rule, and concluded as to the parabolic path of a projected body. A quotation from Galileo's *Dialogue* (Galileo 1630, p. 186), makes clear that he was aware of the invariance of the physical phenomena with respect to velocity transformations:

> Shut yourself up with some friend in the main cabin below decks on some large ship, and have with you there some flies, butterflies, and other small flying animals. Have a large bowl of water with some fish in it; hang up a bottle that empties drop by drop into a wide vessel beneath it. With the ship standing still, observe carefully how the little animals fly with equal speed to all sides of the cabin. The fish swim indifferently in all directions; the drops fall into the vessel beneath; and, in throwing something to your friend, you need throw it no more strongly in one direction than another, the distances being equal; jumping with your feet together, you pass equal spaces in every direction. When you have observed all these things carefully (though there is no doubt that when the ship is standing still everything must happen in this way), have the ship proceed with any speed you like, so long as the motion is uniform and not fluctuating this way and that. You will discover not the least change in all the effects named, nor could you tell from any of them whether the ship was moving or standing still.

Since the independence of the phenomena with respect to time translation, space translation, and orientation was at least implicitly evident at that time, it is perfectly justified that the 10-parameter invariance group of 'Galilean transformations' should carry his name (see also Sect. 1.5).

In the period between Galileo and Newton the dominant figure in physics was surely Huygens (1629–1695). Since much of his work was published only posthumously, some results, usually attributed to Newton, were already anticipated by Huygens. Extremely important technical inventions by Huygens were the precise pendulum clock and the spring clock. (Right up to now, high precision experiments have always been based on precise time-keeping or frequency measurements, currently reaching a precision of 10^{-18}.) He also showed that a pendulum with a frequency independent amplitude has to move on a cycloid. Detailed experiments and analyses with elastic collisions brought him to the conservation laws for momentum and (kinetic) energy, and to an even clearer and more general understanding of the principle of relativity than the one expressed above by Galileo. Possibly Huygens' most important contribution—which led in Newton's hands to the derivation of the general law of gravitational attraction from Kepler's laws for the planetary orbits— was his derivation in 1669 of the correct 'centrifugal' force, a term coined by Huygens: $F_c \sim v^2/r$. (Later, in Chap. 4, the centrifugal force will again play a decisive role in the attempt to derive it as a gravitational force, along the lines of Mach, Einstein, and Thirring.) Huygens also realized that a precise concept of 'force' has to be based on the analysis and measurement of accelerations, and through that he came closer than anybody else to anticipating Newton's second law. However, near the end of his life, when he studied Newton's *Principia*, he refused

Newton's concept of absolute space, and also his concept of gravitation as an action-at-a-distance force.

Naturally, the work of Newton (1643–1727) built decisively on the results of his predecessors, and in particular, Galileo and Huygens. Besides deepening and generalizing their concepts and laws, he also contributed fundamentally new ideas, and eventually created a work which largely shaped the future science of physics—and not only mechanics and gravitation—for the following 200 years. Many of his most innovative discoveries were made and developed when he was still young (in the years 1664–1666), working in isolation at his home village of Woolsthorpe when Cambridge University was closed due to the plague. However, the general and systematic compendium of all his ideas and work, including precise numerical calculations and predictions, was only published in 1687 at the request of his contemporaries and discussion partners Halley, Hooke, and Wren. The resulting *Principia* (Newton 1687) became the most impressive, most important, and most influential physics publication of all time. It comprised a comprehensive treatise of all motions (whether terrestrial or celestial) which could be produced from a mere handful of general principles formulated in a mathematically rigorous framework. Naturally, of central importance for our purposes are his basic laws of mechanics (Newton 1687, pp. 416–417):

Law 1   Every body perseveres in its state of being at rest or of moving uniformly straight forward except insofar as it is compelled to change its state by forces impressed.

Law 2   A change in motion is proportional to the motive force impressed and takes place along the straight line in which that force is impressed.

Law 3   To any action there is always an opposite and equal reaction; in other words, the actions of two bodies upon each other are always equal and always opposite in direction.

In the *Principia*, these laws are preceded by definitions of some physical quantities, e.g., quantity of matter, quantity of motion, impressed force, and centripetal force. Even if, from today's perspective, some of these definitions are no longer tenable, some formulations of the laws are not optimal, and a precise definition of inertial systems is missing, Newton clearly distinguished between definitions and laws, and he also made it explicit that the three laws describe different facts of nature. [This is in contrast to some modern textbooks, e.g., Marion (1965, p. 58), as quoted in the Preface.] Concerning laws 1 and 2, Newton gives much credit to Galileo, and concerning law 3, to Huygens, Wallis, and Wren (Newton 1687, pp. 424–425). However, it is clear, for example, that Newton generalized the concept of 'force' far beyond the work of his forerunners, making it much more precise by combining the phenomena of inertia, collision processes, and centrifugal force. And this concept would later demonstrate its usefulness even beyond Newton's own expectations, in a wide variety of new phenomena and applications. In view of these many different types of 'force', a concise and general definition of the concept seems impossible even today. It may therefore be that we must retreat to the remarkable formulation in Feynman et al. (1965, pp. 12–2): "the law [law 2] is a good program for analyzing nature. It is a suggestion that the forces will be simple". At the end of the nineteenth century there were even (not quite convincing) attempts (e.g., by Kirchhoff, Hertz, and Mach) to completely eliminate the concept of 'force' from physics. As for any concept or law in physics, there are, however, limits to the usefulness of the concept of 'force', e.g., in quantum theory and general relativity.

Besides the three basic laws of mechanics, a second all-important highlight of the *Principia*, and a topic of central importance for us, is the correct law for a universal gravitational force. For a long time, the nearly perfect homogeneity of the Earth's gravitational field on the human and laboratory scale delayed the realization that gravity is in fact a long-range and universal central force field. And it needed Newton's boldness (whether or not it was inspired by an apple falling from a tree in his Woolsthorpe garden) to hypothesize that gravity extends right out to the Moon and the other bodies of the Solar System, and is responsible for the elliptic orbits of all planets, described by Kepler's laws. (Vague conjectures of this sort are already to be found in the work of Kepler, who also coined the term 'inertia'.) It also required Newton's sovereignty in the geometric analysis of conics and his command over the mathematics of infinitesimal elements to derive in detail the $1/r^2$ law of gravitational attraction from the three Kepler laws, viz., elliptic orbits, the area law, and proportionality between the square of the period of revolution and the cube of the major axis of the ellipse. (A more qualitative conjecture of a $1/r^2$ gravitational force was already formulated by Hooke some years before Newton.) The details of Newton's derivation are to be found in Proposition 11 of Book I of the *Principia* (Newton 1687, p. 462), and are explained in more modern terms in, e.g., Cohen (1981) and Barbour (1989, Sect. 10.9). The concept of gravity as an action-at-a-distance force with quasi-infinite range was of course difficult to accept, even for Newton, and much more so for most of his contemporaries, e.g., Huygens and Leibniz. Newton's difficulties with an action-at-a-distance force are wonderfully expressed in a famous quotation from his letter to Bentley (Newton 1756):

> That one body may act upon another at a distance through a *vacuum*, without the mediation of anything else, by and through which their action and force may be conveyed from one to another, is to me so great an absurdity that I believe no man who has in philosophical matters a competent faculty of thinking can ever fall into it.

Today we know from general relativity that gravity as a whole is a local and causal field theory where changes, as far as they contain variable quadrupole moments, propagate maximally with light velocity. However, the conceptual difficulty with the acausal static gravitational field persists to a certain extent: geometrically, we may say that a massive body produces for all time a curved manifold around itself in which all other bodies have to move on predetermined orbits. More mathematically, we know that the Einstein field equations of general relativity also contain, besides the dynamical equations (e.g., for gravitational waves), time-independent constraints which include, in an appropriate approximation, Newton's $1/r^2$ action-at-a-distance law, similar to Coulomb's law in electrodynamics. (We will come back to these questions in Sect. 3.2.) By analyzing in detail the orbits of all planets and moons of the Solar System, and in particular the disturbances of the elliptical orbits by additional masses, Newton could prove that gravity really is a universal force in which all massive bodies in the universe participate. Due to Einstein's law $E = mc^2$, we know today that any physical system which, by definition, has energy and therefore mass, participates actively and passively in gravitation. In his *Principia*, Newton also derived other characteristic and important properties of the gravitational force, e.g., in Newton (1687, p. 590), he showed that, in the region exterior to it, an arbitrary, finite, spherical mass distribution acts like a point mass at its center. In Sect. 1.6, we will give a summary of other specific properties of Newton's law of gravitation.

Finally, we should make some comments on Newton's somewhat surprising concepts of absolute space and absolute time in a so-called Scholium (Newton 1687, pp. 408–415), placed between the definitions and the laws. On the one hand, it is understandable on philosophical if not religious grounds that Newton was also looking for a unique, fixed background (not just the somewhat incomprehensible infinity of equivalent inertial systems) to take as starting point for his physical laws. And a glance at the sky with its so-called fixed stars may have been another motivation for this, as would also have been Newton's opposition to Descartes' relativism. It has also to be said that Newton's formulation of law 1 would lose much of its nontrivial physical content, and would reduce to a kind of definition, if it did not rely (at least implicitly) on the concepts of absolute space and absolute time. On the other hand, Newton's attempt to prove the existence of absolute space by his famous experiment with the rotating bucket (Newton 1687, pp. 412–413) is doomed to failure because this experiment has identical results in all inertial systems, and distinguishes only rotation from linear velocity. It is also quite evident that the concepts of absolute space and absolute time were only a type of makeshift solution for Newton, and throughout most of the *Principia*, he only makes use of relative rather than absolute concepts. Topics like his famous Corollary 5 [the relativity principle, in Newton (1687, p. 423)] can even be seen to contradict the absolute concepts. How cautious, if not reluctant, Newton was about these concepts can also be seen from the following quotations from the *Principia* (Newton 1687, pp. 410–411): "It is possible that there is no uniform motion by which time may have an exact measure", and "For it is possible that there is no body truly at rest to which places and motions may be referred." Newton's self-critical attitude is wonderfully described in Einstein (1927):

Newton himself was better aware of the weaknesses inherent in his intellectual edifice than the generations of learned scientists which followed him. This fact has always aroused my deep admiration.

Einstein also formulated with admirable clarity the reason why concepts like absolute space and absolute time should not have a place in physics (Einstein 1922):

It is contrary to the scientific mode of understanding to postulate a thing which acts, but which cannot be acted upon.

It is well known that, even in Newton's day, there were prominent opponents to Newton's absolute concepts, in particular Huygens, Leibniz, and Berkeley.

However, it took nearly 200 years to arrive at clear and consistent formulations of Newton's laws, especially law 1, which avoided or eliminated any recourse to absolute space and absolute time, as we will explicate in Sect. 1.2.

1.2   The Work of Ludwig Lange

Before coming to the work of Ludwig Lange, we should comment on Neumann's inaugural address (Neumann 1870) at the University of Leipzig, which is a kind of forerunner of Lange's work and which is quoted by Lange as a decisive stimulus for his own work. Neumann manages to replace Newton's absolute time by an operational definition: he assumes, without stating this explicitly, that force-free particles exist and can be identified as such, and also that absolute space can be observed directly. He then abstracts from experimental observation that "two material points, each force-free, move in such a manner that equal path distances of the one always correspond to equal path distances of the other". [A similar formulation can also be found in the later treatise (Thomson and Tait 1879, pp. 246–248)]. From this he concludes that "equal time intervals can be defined as those in which a force-free particle moves equal space intervals". (Definition of an inertial clock!) It is somewhat strange that Neumann does not make any attempt along similar lines to replace absolute space by an operational definition. On the contrary, he strengthens Newton's concept of absolute space, and tries to make it more concrete by saying that "at some unknown place in the universe there is an unknown body, indeed an absolutely rigid body, a body whose shape and size are unchanged for all time", and he calls this 'Body Alpha'. [To some extent this body was already anticipated by Newton when he says in Newton (1687, p. 411), "since it is possible that some body in the regions of the fixed stars or far beyond is absolutely at rest".] However, the concept of this Body Alpha loses much of its privileged status because Neumann has to admit that there would exist infinitely many equivalent bodies due to Galileo's principle of relativity.

In the years 1885–1886, L. Lange (at the age of just 22–23!) published three papers on the law of inertia which must certainly be judged as a stroke of genius because, for the first time, and nearly 200 years after Newton's *Principia*, he succeeded in eliminating the concepts of absolute space and absolute time, replacing them by a clear operational analysis in which he determines which parts of Newton's law 1 are useful definitions and which parts are non-trivial experimental facts, if not wonders of nature. The papers Lange (1885a, 1886) are quite extended. Besides Lange's innovative contributions, they also include historical and philosophical overviews of the law of inertia, while Lange (1886) contained Lange's Ph.D. thesis, presented in 1886 at the University of Leipzig. The shorter paper (Lange 1885b) concentrates on Lange's own contributions and their mathematical foundations. In the following, we mainly refer to and quote from this paper, for which an English translation has recently been published, together with a commentary (Pfister 2014).

In 1902, Lange published another quite extended analysis (Lange 1902) of the law of inertia, together with reactions to his earlier work by other authors (see below).

The motivation for Lange's work and its main results are clearly expressed by the following quotation from Lange (1885b):

> Newton's absolute space is a phantom that should never be made the basis of an exact science. [...] To find a fully valid substitute for it is the goal of the following.

As already mentioned, in the elimination of Newton's absolute time, Lange follows the results of Neumann (1870) when he says in Lange (1885b):

> The fundamental timescale of dynamics is to be defined through the motion of a point left to itself. Under this viewpoint, the law of the uniform motion of all points left to themselves is, as Thomson and Tait (1879) correctly note, a pure convention for one such point, and it is more than convention, it is a research result, only insofar as it applies to any other points left to themselves. [...] The question now arises whether it is possible to eliminate also the absolute space by a similar procedure. Indeed this is possible. The fundamental coordinate system of dynamics may be called 'inertial system', the fundamental timescale of dynamics 'inertial timescale'.

This is clearly the first place where the now standard terms 'inertial system' and 'inertial timescale' are introduced into the literature.

> In exactly the same way as the one-dimensional inertial timescale could be defined through one single point left to itself, the three-dimensional inertial system can be defined through three points left to themselves.

[Originally, in Lange (1885a), Lange tried to base an inertial system on only two points left to themselves. After an objection by the mathematician A. Voss, he corrected this in an addendum on pp. 539–545 of Lange (1885a) to three points left to themselves.]

> Herefrom follows that the law of the constant direction of motion of force-free points is pure convention for three such points, but embodies a noteworthy research result insofar as it is valid for more than three, for arbitrarily many points in relation to one and the same system.

Lange then summarizes his results in an admirably clear and concise way:

Definition I. 'Inertial system' is called any coordinate system of the kind that in relation to it three points P, P', P'', projected from the same space point and then left to themselves—which, however, may not lie in one straight line—move on three arbitrary straight lines G, G', G'' (e.g., on the coordinate axes) that meet at one point.

Theorem I. *In relation to an inertial system the path of an arbitrary fourth point, left to itself, is likewise rectilinear.*

Definition II. 'Inertial timescale' is called any timescale in relation to which one point, left to itself (e.g., P), moves uniformly with respect to an inertial system.

Theorem II. *In relation to an inertial timescale any other point, left to itself, moves uniformly in its inertial path.*

There is an interesting footnote to Theorem II:

One sees that, in a certain sense, the space and the time part of the law express the same fact twice, only in one case with respect to the three-dimensional space, in the other with respect to the one-dimensional time.

This can be seen as a first hint that in the four-dimensional language of special relativity the two definitions and the two laws do indeed combine to form one definition and one law. It is therefore natural that, in our further analysis of the law of inertia in Sects. 1.3–1.6, we mainly use this four-dimensional language. [From a more philosophical viewpoint, such a procedure is particularly advocated in Earman and Friedman (1973).] In any case, it is remarkable that the law of inertia is one of the few laws which read identically in nonrelativistic and relativistic physics. Moreover, although inertial systems were first defined and applied in mechanics, it turned out later that in the same systems, the electromagnetic, optical, thermodynamical, and quantum-mechanical laws also take their simplest form, which can be seen as a convincing sign of the unity of all of physics.

Part II of Lange's paper (Lange 1885b) provides a detailed proof that the Definition I of an inertial system is indeed mathematically consistent, a fact which appears today quite evident on the basis of linear algebra. Somewhat superfluous, Lange aggravates this analysis by sticking to orthogonal coordinate systems, whereas the projective structure of spacetime would be quite adequate for an analysis of the law of inertia, as we will explicate in Sect. 1.4. In his part II, Lange also analyzes the whole manifold of equivalent inertial systems which results from the Galilean relativity principle. Generally speaking, part II of the paper (filling nine pages in the original) is quite circumstantial and awkward, containing many unnecessary repetitions, and also important loopholes and omissions, so that one may agree with J. Barbour who, in a recent article (Barbour 2004), accused Lange's construction of being "rather awkward and clumsy".

The important content of Lange's part II has recently been elegantly extracted (in a two-page appendix) in Giulini (2002): it can be proven that, in three-dimensional space locally (for fixed time), the positions and the directions of three (arbitrary) paths (but not more than three paths!) can be transformed to prescribed new positions and directions such that the transformed ones constitute three straight lines. What is missing in the work of Lange and Giulini are some global conditions on the three (arbitrary?) paths: it is intuitively clear that these can only be transformed in a non-singular and non-degenerate way (mathematically, by a global diffeomorphism) to straight lines if the original paths are smooth, if they are not closed, i.e., they extend at both ends to infinity, if any two of these paths have at most one meeting point, and presumably also if some other conditions are fulfilled.

We think that a much simpler proof that Lange's Definition I is mathematically consistent can proceed in the following way. One starts from three straight lines along axes x, y, z which have to be linearly independent but not necessarily orthogonal. Application of a global diffeomorphism

$$\xi = f_1(x, y, z) , \qquad \eta = f_2(x, y, z) , \qquad \zeta = f_3(x, y, z) ,$$

with largely arbitrary differentiable functions f_i, transforms these straight axes to new paths which are nearly arbitrary but fulfil the global conditions stated above. The inverse transformation, which exists and is likewise non-singular and non-degenerate, then realizes the requirement formulated by Lange in his Definition I. Furthermore this recipe is literally extensible to the four-dimensional space-time in order to cover the space and time part of the law of inertia. For simplicity, Lange and Giulini confined themselves to "three points P, P', P'' projected from the same space point". It should, however, be clear that three paths, of which two at a time are skew to each other, can also serve as the basis of an inertial system.

In part III of Lange (1885b), Lange reviews the work of other authors who, shortly before Lange, also critically analyzed the foundations of mechanics, in particular Newton's concepts of absolute space and time. He remarks that this activity "can serve as further evidence that the question dealt with has presently become a vital one". In this connection one may also refer to the quite voluminous historical work (163 pages in total) by E. Wohlwill on *The discovery of the law of inertia* (Wohlwill 1883, 1884), not quoted by Lange. We have already mentioned the papers Neumann (1870) and Thomson and Tait (1879). Lange also refers sympathetically to the famous and influential books (Mach 1872, 1883) by E. Mach who was the most explicit critic of Newton's absolute space, and who gave convincing arguments for relating inertia to the overall distribution of matter in the universe. (We will come back to some of these arguments in Sect. 4.1.) However, Mach did not give arguments showing how to define the local inertial systems without recourse to absolute space. This was attempted in the book (Streintz 1883) by H. Streintz, a student of C. Neumann. He tried to base an inertial system on the axes of rotating gyroscopes, something which should be possible in principle. However, since he did not clearly discriminate between definitions and laws, he was in danger of running around a methodological circle, as Lange correctly pointed out. Finally, Lange quotes the paper (Thomson 1883a) by J. Thomson, the elder brother of W. Thomson (Lord Kelvin). He, like Lange, based the inertial systems on the paths of force-free particles. However, Thomson once again did not distinguish between definitions and laws. And what he called his law of inertia was, if anything, only a poor formulation of Newton's second law, while the last sentence of the paper suggests that Newton's first law is a special case of his second law, which is of course fundamentally wrong. More interesting is a *Note on Reference Frames* by P.G. Tait which is added to a second (not very illuminating) paper (Thomson 1883b, pp. 743–745) by J. Thomson. Here, Tait presents an alternative proof that Lange's Definition I is mathematically consistent, one which is much shorter and clearer than Lange's own derivation. It is based on a number of 'snapshots' of the paths P, P', P'', and the analysis of their relative distances. (One may criticize that the projective structure suffices to formulate the law of inertia, and a metric structure is not really necessary.) A reanalysis of Tait's nice construction in modern mathematical terms can be found in Giulini (2002, 2015) .

In the rest of this section we would like to document (mostly in chronological order) the fact that, in the 25 years after publication of the Lange papers, this work was frequently, prominently, and almost invariably positively reviewed. Already in the year 1886, A. König reported in detail about this new and "uncontradicted foundation of mechanics" to the Physical Society of Berlin (König 1886). In Seeliger (1887), the well-known astronomer H. Seeliger gave an eight-page report on Lange's papers (Lange 1885b, 1886):

> These will constitute a very important contribution to the clarification of the fundamental principles of the theory of motion, and will earn the attention of all people who are generally interested in the analysis of fundamental concepts. [...] The author [Lange] has succeeded in an excellent and surprisingly enlightening way [to clear up the foundations of mechanics] by stating the following definitions and laws.

Here Seeliger quotes Lange's Definitions I and II and the Laws I and II verbatim (see above). Later, in the year 1906, Seeliger repeated many of these positive judgements of Lange's work in a more extended article *On the so-called absolute motion* (Seeliger 1906). In his remarkable article *Things at rest in the universe* (Schwarzschild 1897), Seeliger's Ph.D. student K. Schwarzschild chose a formulation of the law of inertia which was almost identical with Lange's (although he did not cite Lange).

The most prominent appreciation of Lange's work came from E. Mach who, in the second edition (1888) of his book (Mach 1883), wrote:

> Lange's paper (Lange 1886) seems to belong to the best of what has been worked out regarding these questions. I have much sympathy for the methodological approach. The careful analysis and the historical–critical consideration of the concept of motion has, it seems to me, brought results of lasting value.

In the seventh edition (1912) of Mach (1883), Mach even states in his preface: "I thank Mr. L. Lange and Mr. J. Petzoldt not only for their agreement in details but also for their active and successful cooperation." And in the preface of the ninth edition (1933) Ernst Mach's son Ludwig speaks of Mach's "correspondence particularly with Ludwig Lange".

A very interesting analysis of Lange's paper (Lange 1886) from the logical and linguistic standpoint is provided by the well-known logician G. Frege in Frege (1891). On the one hand, he gives high praise:

> The paper warrants reading and is suitable to dispel false security [about the law of inertia], and to stimulate further reflection. [...] It is no humble merit of the author to have substituted a hidden relation by a clearly expressed one.

On the other hand Frege criticizes the fact that Lange leaves open the question as to what "a particle left to itself" (or a force-free particle) might be, and what "uniform motion" might mean, if one cannot really compare lengths at different times. In his later review (Lange 1902), Lange admits that his work does sometimes fall short in this regard. In Sects. 1.3 and 1.4, we shall try to make some progress on these intricate questions by Frege. Frege's paper is really fun to read, and it "stimulates further reflection", e.g., by sentences like "If one cannot answer a question, one can at least make it disappear behind a cloud of imprecise statements".

In the article (Voss 1901) by A. Voss about the principles of rational mechanics, which appeared in the *Encyclopedia of the Mathematical Sciences, Including*

Their Applications, the author states: "Lange has tried to remove this deficiency [concerning the reference systems of mechanics] by specifying a system which at least does not possess any logical or methodological defect." A quite extended analysis (filling 17 pages) of Lange's work is given by J. Petzoldt in Petzoldt (1908). On the one hand he states: "The most penetrating and most far-reaching attempt in this direction [to save Newton's mechanics], and which has also met with the most approval, is the one by Ludwig Lange." On the other hand, Petzoldt formulates a pedantic and in parts untenable critique of the details of Lange's construction.

Finally, we may refer to M. von Laue's book on relativity theory, Vol. 1 (von Laue 1911) (the very first textbook on special relativity!), in which he presents Lange's Definitions I and II and his Laws I and II verbatim, and assesses them with the words: "To have set these as substitutes for the rightly called 'somewhat ghostly' absolute space and absolute time of Newton, is the great deed of Ludwig Lange." Later, in the year 1948, von Laue published a sort of biography (von Laue 1948) of Lange with the title *Dr. Ludwig Lange, 1863–1936. (An unjustly forgotten person.)* We learn there that, from 1887 on, Lange suffered from a mental disorder that required him at times to frequent mental clinics. He continued to publish on many different topics in his later years (not just physics, but also psychology, photochemistry, and calendar science), but his results never reached the quality of his work in the Ph.D. period, and he never obtained a good position at a German university.

Today, Lange seems to be forgotten, not only as a person, but also for his "great deed" of clarifying the real substance of the law of inertia. The only more recent textbook we know of, which quotes and appreciates Lange's work, is *Relativity and Geometry* by R. Torretti (1983). We would like to argue, however, that nearly every modern textbook on mechanics would be considerably improved if its often inconsistent, circular, or otherwise untenable formulations of the law of inertia were replaced by a verbatim quotation of Lange's Definitions I and II, and his Laws I and II. (Compare our criticism of mechanics textbooks in general, and of one of them in particular, in the Preface.)

1.3 What is a Free Particle?

The law of inertia is usually based on so-called *free particles*, or, in the words of L. Lange, particles left to themselves, sometimes also called force-free particles. But it is seldom explained how such particles can be precisely defined and realized experimentally. Picking out free particles by saying that they are free of external forces—notwithstanding the fact that such a characterization already appears in Newton's *Principia* (Newton 1687, p. 416) and in many modern textbooks—is logically fallacious: the definition of the terms 'force' and 'force-free' form part of Newton's second law, which itself relies on the inertial systems defined in the first law. In addition, the popular characterization of free particles as those that are far from any other objects cannot be accepted because it is impossible in any practical case to say what distance from other objects would be big enough, especially since gravity and electromagnetism have (in principle) an infinitely long range.

In contrast to these characterizations by *external* conditions we try in the following to present a somewhat new definition of free particles by *internal* properties. For this purpose, we start from a very general class of 'objects' in nature, and reduce or specialize in consecutive steps to ever simpler objects, until we reach so-called *inactive test objects* with no other physical properties than mass (compare Pfister 2004). It is often not sufficiently appreciated how kind nature is in supplying us with 'objects', i.e., with subsystems of the universe which possess characteristic properties (literally in the sense 'proper to the system') that can be described and measured almost without recourse to the rest of the universe, and that these objects are localized in space, but 'live' for an almost infinitely long time. And usually the same types of objects (with the same or very similar properties) can be found or produced at any place and at any time. (Obviously there are properties of objects like colour, temperature and hardness, which have little influence on their motion, and which are therefore left out of consideration.)

As a first simplification in the class of all objects appearing in nature, a reduction to inactive objects is surely recommended, i.e., to objects having constant inner properties, and having no 'inner motor' (in contrast to humans, robots, cars, airplanes, etc.), and not disintegrating during their 'life' (in contrast to supernovae, radioactive particles, etc.). Furthermore, a reduction to 'test objects' suggests itself, i.e., to objects having (ideally) no back-reaction on the rest of the universe. In practice, small objects are the best candidates for this requirement but, depending on the circumstances and the questions we ask, planets and stars can also function as inactive test objects. It is, however, important to stress that we are here concerned only with classical physics, i.e., we must not take the limit to the even 'smaller' elementary particles with their quantum mechanical and non-local entanglement properties.

Experience with nature tells us that such inactive test objects move on uniquely predictable paths (one-dimensional submanifolds of the four-dimensional space-time), i.e., for a given object with given properties, the complete path is—under given exterior circumstances, i.e., given 'physical fields'—fixed through an 'initial' event, an initial direction (spatial direction and velocity) at this event, and possibly other initial data (see below). (Mathematically, the dynamics of these objects is ruled by second order differential equations.) However, in nature there is a great variety of different classes of inactive test objects which, starting from the same event with the same initial direction, follow in the future completely different paths, depending on properties like charge-to-mass ratio, higher mass and charge multipoles, and possibly other distinguishing properties.

We propose now to characterize the free particles by demanding that they have as few nontrivial properties as possible. They should have charge $q = 0$, magnetic moment $\boldsymbol{\mu} = 0$, all higher electromagnetic multipole moments $q_{ijk...} = 0$, intrinsic angular momentum $\mathbf{j} = 0$, and all higher mass multipole moments $m_{ijk...} = 0$. Furthermore, (almost) every other physical property imaginable, or for which experimentalists have invented a measuring device, should also be zero. This restriction certainly brings about a great simplification because the path of such an object is indeed determined by the initial event and direction, without giving additional initial data. (In contrast, for objects with nontrivial higher moments, one has to give not only the values of these moments but also their orientation relative to the initial path direction.) The only physical property of the free particle that has to be nonzero is the mass m, or the rest mass m_0, because otherwise the manifold of paths through an initial event would be restricted to the three-dimensional manifold of lightlike paths.

In this way, we arrive at the following:

Definition 1.1. Free particles are inactive test objects with only one nontrivial physical property: mass.

These free particles and their paths in spacetime are now distinguished by some 'wonders of nature' which unfortunately do not come to life in most textbooks:

- Nature provides these nearly featureless objects in a universal manner everywhere and at any time.
- Their paths are independent of the mass value and of many other remaining 'inner properties' like chemical constitution. This fact is usually formulated as 'all bodies fall alike', or 'inertial and gravitational mass are equal', or as part of Einstein's equivalence principle. But it is rarely stressed that this property is already essential for the formulation of Newton's first law, and for the possibility of endowing spacetime with a simple and unique inertial structure.
- Free particles are the only objects whose paths can (in the absence of gravity) function as axes of a unique *global* coordinate system (in the sense of Lange): free particles, emanating from an event in different spacetime directions, do not meet again, in contrast, e.g., to the focusing of charged particle paths in appropriate electromagnetic fields. In strong gravitational fields, even free particles can be focused (by gravitational lenses), with the consequence that no really global inertial systems exist, another fact which is mentioned rather seldom in mechanics textbooks.

Some of the above arguments, in particular our definition of the term 'free particle' [as they appeared also in Pfister (2004)] were criticized in a recent book (Brown 2005) by H. Brown. He asks the question whether it "does not rely too much on hindsight—whether indeed the definition of the properties that are supposed to have a null value does not ultimately refer to the very inertial frames we are trying to construct". Of course, any of the experiments leading to the classification described above, will rely on *some* presuppositions, e.g., on available materials. And if nature functioned in a totally different way, they might not be possible, or not lead to the desired results. However, we would like to argue that the existence and the precise properties of inertial systems are not a prerequisite for our procedure for defining free particles, as is evident from the fact that concrete laboratory systems are never exact inertial systems but are exposed to the Earth's gravity and the Earth's rotation. We should also say that a private discussion between H.P. and H. Brown (at the conference "Beyond Einstein" in Mainz, Germany, September 2008) resulted in the agreement that at least some of the critical remarks in Brown (2005) concerning the paper (Pfister 2004) are based on misunderstandings, and are not maintained in their sharp form.

1.4   A New Definition of a Straight Line, Connections with Projective Geometry, and Newton's Law of Inertia

The term 'straight line' is not so innocent and evident as one may naively think. It is usually based on the technical concept of a rigid rod. But this is surely unsatisfactory if one is interested in the foundations of mechanics, or of the whole of physics, because rigid bodies are complicated secondary concepts of mechanics (if not of quantum mechanics). Furthermore, in celestial mechanics, which was historically the midwife and the first testing ground for Newtonian mechanics, a straight line can never be realized by a rigid rod, and in special and general relativity, the usual concept of a rigid body breaks down anyway.

Instead we distinguish here the straight lines in the host of all path structures by characteristic relations between their elements. This will be done first in the usual coordinate-dependent formalism, and afterwards in a (new) purely geometric and coordinate-independent language. And we will formulate all arguments in the arena of four-dimensional spacetime, whose 'points' represent events, i.e., the most elementary and idealized experimental facts, like the collision of two 'point-particles'. In technical, mathematical terms, spacetime $\mathcal{M}$ should be a connected, paracompact Hausdorff manifold of dimension four, and of differentiability class $C^k (k \geq 3)$, and 'paths' should be one-dimensional submanifolds of $\mathcal{M}$, also of differentiability class $C^k (k \geq 3)$. In an arbitrary coordinate system $\langle x^\mu \rangle$ in $\mathcal{M}$, and with an arbitrary monotone parameter τ on each path $x^\mu(\tau)$, the property that the paths of inactive test objects are uniquely determined by an 'initial' event $x^\mu(\tau = 0)$ and an initial direction $dx^\mu/d\tau(\tau = 0) = v^\mu(0)$ in this event has the consequence that the higher derivatives $d^i x^\mu/d\tau^i (\tau = 0), (i \geq 2)$ are already fixed by $x^\mu(0)$ and $v^\mu(0)$, or that one generally has

$$d^2 x^\mu/d\tau^2 = f^\mu(x^\nu, dx^\lambda/d\tau) , \tag{1.1}$$

with given, sufficiently differentiable functions f^μ. A mathematical distinction of a special path structure is then established by the choice of special functions f^μ, and surely the simplest one is the choice f^μ identically zero. Then, the integration of (1.1) immediately gives $x^\mu(\tau) = x^\mu(0) + \tau v^\mu(0)$, i.e., an ensemble of straight lines. However, this characterization is not invariant under (nonlinear) coordinate and parameter transformations, a defect that was masterfully denounced in Einstein (1920):

> I quote Galileo's law of inertia as an example [of the mixing of statements about the means of description and statements about the object to be described]. It reads in detailed

formulation necessarily as follows: matter points that are sufficiently separated from each other move uniformly in a straight line—provided that the motion is related to a suitably moving coordinate system and that the time is suitably defined. Who does not feel the painfulness of such a formulation? But omitting the postscript would imply a dishonesty.

Applying a general coordinate transformation $x^\mu \longrightarrow x'^\mu(x^\nu)$ and a parameter transformation $\tau \longrightarrow \sigma(\tau)$ to $d^2x^\mu/d\tau^2 \equiv 0$ leads to

$$\frac{d^2x'^\mu}{d\sigma^2} + \Pi^\mu_{\nu\lambda}(x'^\rho)\frac{dx'^\nu}{d\sigma}\frac{dx'^\lambda}{d\sigma} = k(\sigma)\frac{dx'^\mu}{d\sigma} \ , \tag{1.2}$$

with the symmetric 'projective coefficients'

$$\Pi^\mu_{\nu\lambda} = -\frac{\partial^2 x'^\mu}{\partial x^\rho \partial x^\sigma}\frac{\partial x^\rho}{\partial x'^\nu}\frac{\partial x^\sigma}{\partial x'^\lambda} \ , \tag{1.3}$$

and the factor $k(\sigma) = (d^2\tau/d\sigma^2)(d\sigma/d\tau)$. The physical meaning of (1.2) is eloquently characterized by another quote from Einstein (1922, pp. 81–82):

> The unity of inertia and gravitation is formally expressed by the fact that the whole left-hand side of this equation has the character of a tensor (with respect to any transformation of coordinates), but the two terms taken separately do not have tensor character. In analogy with Newton's equations, the first term would be regarded as the expression for inertia, and the second as the expression for the gravitational field.

Comparing (1.2) with the equation $d^2x^\mu/d\tau^2 \equiv 0$ in the original coordinates and with the original parameter, one can also say that (four-dimensional Minkowski) forces which are proportional to the four-velocity $v'^\mu = dx'^\mu/d\sigma$ or are symmetrically bilinear in this variable (with appropriate coefficients) can be 'transformed away'. In this connection it may be mentioned that the simplest Minkowski force K^μ which cannot be transformed away and which, through

$$0 = \frac{d}{d\tau}c^2 = \frac{d}{d\tau}\left(\frac{dx^\mu}{d\tau}\frac{dx_\mu}{d\tau}\right) \sim K^\mu v_\mu \tag{1.4}$$

is compatible with the constancy of the light velocity, is given by $K^\mu \sim F^{\mu\nu}v_\nu$, with an antisymmetric tensor $F^{\mu\nu}$. This is therefore the Lorentz force law of electrodynamics. (We are not aware that this remarkable characterization of the Lorentz force appears anywhere in the literature.)

As announced, we will now provide an alternative characterization of straight lines in purely geometric terms in order to avoid the 'painfulness' of the above coordinate-dependent characterization. The safest way to accomplish this goal is a reduction to the most primitive, experimentally decidable facts, the events. And indeed, historically one of the first elements of an exact natural science (of celestial mechanics) was provided by the observations of the events of planetary occultations, long before the invention of coordinate systems. On the theoretical side, it was E. Kretschmann in Kretschmann (1915) who first suggested such a reduction, as a reaction to Einstein's wavering between only partly and fully covariant formulations of a relativistic theory of gravity. And it was in a letter to P. Ehrenfest on 26 December 1915 (Schulmann et al. 1998, Doc. 173) that Einstein himself gave the following formulation:

> The physically real in the world of events (in contrast to that which is dependent upon the choice of a reference system) consists in spatiotemporal coincidences (and nothing else!). Real are, e.g., the intersections of two different worldlines, or the statement that they do not intersect.

What then is the minimal number of particle paths, and of events of intersection between them, that allows one to uniquely characterize the straight lines, as distinguished from other path structures? For one path (e.g., path P_7 in Fig. 1.1), in order to discriminate between 'straight' and 'curved', one needs at least three points (events) on it. In order for these points to help formulate some characteristic (lawful) properties of paths, the points have to be defined not only by the inevitable crossing of two paths (e.g., in two dimensions) but by the crossing of at least three paths. (Duality between paths and points!) The simplest 'incidence figure' of this type is easily seen to consist of seven paths and seven points, and it is known in geometry as the simplest example of a finite projective plane, the so-called Fano configuration (see, e.g., Stevenson 1972, pp. 22–25). However, a realization of such a figure is possible only in an abstract 'plane' over a field of characteristic two, and not in the continuous and ordered 'physical' plane. In the representation of the Fano figure in the continuous plane shown in Fig. 1.1, at least one path has to be curved.

A similar situation occurs for figures with eight paths and eight points. The simplest generic incidence figure that is realizable in a continuous two- or higher-dimensional space(-time) consists of nine paths and nine points and has been known since ancient times as the Pappus figure. It also represents one of the basic axioms of modern projective geometry. (See any standard textbook on projective geometry, e.g., Kadison and Kromann 1996.) And indeed, a path structure (in two dimensions) can satisfy the Pappus incidence properties only if it consists of straight lines, as was first proven in Hilbert (1899). However, here we prefer to consider in detail the Desargues figure (consisting of ten paths and ten points), because it plays (in some

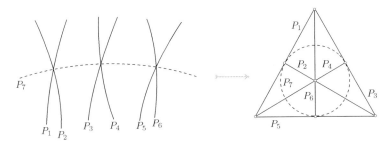

Fig. 1.1 Illustrating the fact that at least seven paths are needed to characterize an invariant property of a path structure. 'Closing' this sketch leads to the standard Fano figure with seven paths and seven points

Fig. 1.2 The Desargues
figure for free particle paths
in a flat manifold (but without
an optimally adapted axis
system) or in a *curved*
manifold (with gravity)

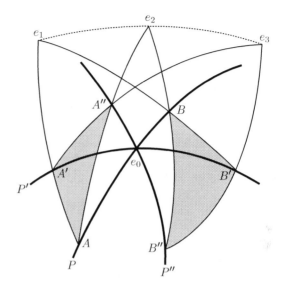

respects) an even more fundamental role in projective geometry, and because it is
not confined to two dimensions (see Fig. 1.2).

Three paths P, P', P'' of the considered path structure emanate from the event e_0
in linearly dependent or independent directions. On P one chooses nearly arbitrary
points A, B, equally A', B' on P', and A'', B'' on P''. The point e_1 is the crossing
point of the paths AA' and BB' in the 'plane' PP', and e_2, e_3 are constructed
equivalently. Then the Desargues theorem says that the path $e_1e_2e_3$ is also a member
of the considered path structure. (In brief geometric terminology, one can say this:
if two triangles $AA'A''$ and $BB'B''$ are in perspective from the point e_0, then they are
also in perspective from the line $e_1e_2e_3$.) As we show below, it is an easy exercise
to confirm that any ensemble of straight lines satisfies the Desargues incidence
properties. We have nevertheless drawn Fig. 1.2 with curved lines because we do
not assume that an optimally adapted system of axes has already been chosen, and
because Fig. 1.2 shall also serve in Sect. 2.3 for a curved manifold.

If the points $A, A', \ldots, B''$ are defined by the vectors $\mathbf{A}, \mathbf{A}', \ldots, \mathbf{B}''$ (in some
coordinate system), the center of perspective e_0 can be represented as $\mathbf{e}_0 = \lambda \mathbf{A} +
(1-\lambda)\mathbf{B}$, or as $\mathbf{e}_0 = \lambda'\mathbf{A}'+(1-\lambda')\mathbf{B}'$, or as $\mathbf{e}_0 = \lambda''\mathbf{A}''+(1-\lambda'')\mathbf{B}''$, with appropriate
coefficients $\lambda, \lambda', \lambda''$, fulfilling, however, $\lambda \neq \lambda' \neq \lambda'' \neq \lambda$. Subtraction of the first
two equations, and division by $\lambda - \lambda' \neq 0$ leads to

$$\frac{\lambda}{\lambda - \lambda'}\mathbf{A} - \frac{\lambda'}{\lambda - \lambda'}\mathbf{A}' = \frac{1 - \lambda'}{\lambda - \lambda'}\mathbf{B}' - \frac{1 - \lambda}{\lambda - \lambda'}\mathbf{B} = \mathbf{e}_1 \ .$$

Since the coefficients on the left side and on the right side of this equation each sum
to the value 1, this vector lies on AA' and BB', and is therefore their crossing point
e_1. Equivalently, we have

$$\frac{\lambda}{\lambda - \lambda''}\mathbf{A} - \frac{\lambda''}{\lambda - \lambda''}\mathbf{A}'' = \frac{1 - \lambda''}{\lambda - \lambda''}\mathbf{B}'' - \frac{1 - \lambda}{\lambda - \lambda''}\mathbf{B} = \mathbf{e}_2 ,$$

$$\frac{\lambda'}{\lambda' - \lambda''}\mathbf{A}' - \frac{\lambda''}{\lambda' - \lambda''}\mathbf{A}'' = \frac{1 - \lambda''}{\lambda' - \lambda''}\mathbf{B}'' - \frac{1 - \lambda'}{\lambda' - \lambda''}\mathbf{B}' = \mathbf{e}_3 .$$

Herefrom follows

$$\mathbf{e}_3 = \frac{\lambda' - \lambda}{\lambda' - \lambda''}\mathbf{e}_1 + \frac{\lambda - \lambda''}{\lambda' - \lambda''}\mathbf{e}_2 ,$$

and since again the coefficients sum to 1 in each part, we see that e_3 lies on the straight line connecting e_1 and e_2.

Conversely, in his *Grundlagen der Geometrie* (Hilbert 1899), D. Hilbert succeeded in building a coordinatization (named 'Streckenrechnung') on the geometry of the Desargues figure (together with the parallel axiom). He showed that this coordinatization satisfies the axioms of the real numbers and that, in these coordinates, the original paths satisfy linear equations. In this way, straight lines can be uniquely characterized, in a coordinate-independent manner, solely by projective geometry. [In this connection, we would like to mention that, even in 1885, L. Lange made the following remark (Lange 1885a): "Concerning the elegance of the systematic, the field of mechanics may take an example from projective geometry." And as H. Weyl put it in Weyl (1931): "From inertia one can directly read off only the projective quality, not the affine connection." See also Sect. 1.5.]

Definition 1.2. Straight lines are an ensemble of paths that fulfil the Desargues property.

It is, however, worth mentioning that, if one enriches the manifold in which the Desargues figure lives by a Minkowski metric, one can arrange the construction in such a way that:

- either all ten paths are timelike, i.e., they can be represented by (free) particles,
- or the 'first' nine paths are timelike, but the path $e_1e_2e_3$ is spacelike, i.e., spacelike straight lines (and hyperplanes) can be constructed from timelike paths, without relying on light rays, as in Einstein's construction of a hyperplane of simultaneity,
- or the triangle sides AA' and BB' are parallel light rays in the plane PP', equally AA'' and BB'' in the plane PP'' (whence the path $e_1e_2e_3$ lies at infinity), with the consequence that also the, e.g., timelike lines $A'A''$ and $B'B''$ are parallel, and light reflected between them can establish a so-called geometrodynamical clock, as first constructed in Marzke and Wheeler (1964), and then simplified in the direction of our above construction in Castagnino (1968).

The essential content of Newton's first law is now the following 'wonder of nature', or in Leibniz' terms the following 'preestablished harmony':

Newton's First Law. *The simplest and most elementary objects of nature, the free particles, move (in the absence of gravity) on the mathematically simplest paths in spacetime, the straight lines.*

In detail, the first four paths of such a construction define an inertial system. However, the fact that all other free particles also move on straight lines with respect to this inertial system, and do so independently of the mass and many other internal properties of the particles, is one of the most fundamental and most marvellous facts of nature. This is also a wonderful example [in the words of E. Wigner in Wigner (1960)] of "the unreasonable effectiveness of mathematics in the natural sciences".

1.5   Spacetime Structures of Newtonian Physics

In the preceding sections we have clarified in depth the foundation and content of the law of inertia in Newtonian physics. Free particles and their paths single out a preferred class of motions which are described relative to some coordinate system $\langle x^\mu \rangle$ and within some given geometry, called spacetime $\mathcal{M}$. As Ehlers observed in the introduction to his talk at the Trieste conference for Dirac's seventieth birthday (Ehlers 1973a, p. 71):

> Space and time and the even more basic notion of spacetime, and the structures assigned to them, belong to the most fundamental concepts of science. So far, every physical theory of some generality and scope, whether it is a classical or a quantum theory, a particle or a field theory, presupposes for the formulation of its laws and for its interpretation some spacetime geometry, and the choice of this geometry predetermines to some extent the laws which are supposed to govern the behaviour of matter, the laws of primary concern to physics.

In this section we show how this preferred set of paths or 'standard motions' are related to various classical, i.e., non-relativistic spacetime geometries, and how they define their affine structure. Although we have emphasized the purely projective nature of the law of inertia, giving $\mathcal{M}$ the structure of a *path space*, we will introduce the notion of a 'straight line' by means of an affine connection and a derivative operator, independently of any metric. [As Giulini remarked in Giulini (2009, Sect. 2.5), the affine group—the group of affine transformations preserving straight lines—already emerges as an automorphism group of inertial structures (determined by the law of inertia) without a privileged parametrisation of spacetime paths, which is usually done by elementary clocks.] This affine point of view was put forward especially by Weyl (1919) and later on by Schrödinger (1950).

The presentation given here largely follows the masterful exposition given by J. Ehlers in Ehlers (1973a,b) and the in-depth analysis of the foundations of spacetime by M. Friedman in Earman and Friedman (1973) and Friedman (1983), which originated from the pioneering works of Weyl (1919, Sect. 20), Cartan (1923), and Friedrichs (1928) in the early stages of general relativity, followed in the 1960s by Havas (1964, 1967) and Trautman (1965, 1967). [See also the exposition given in Misner et al. (1973, Chap. 12), and in particular, Boxes 12.2 and 12.3.] In addition, this section serves as a kind of preparation for the deduction of the different levels of geometrical structures of our 'world' within Einstein's general relativity in the next chapter.

To explore the nature of space and time, and of spacetime, we use mathematical methods of differential geometry. For the main part, we describe these classical spacetime theories in the four-dimensional, geometrical language of special relativity. There are some convincing arguments for such an approach, unusual to most textbook presentations of classical mechanics, which are based on the standard 3-vector calculus on some three-dimensional spacelike sections of the four-dimensional 'world'. On the mathematical side, a detailed comparison of Newtonian concepts of space and time with the spacetime structures underlying the relativistic theories of special and general relativity (which are most naturally expressed in terms of concepts of differential geometry, and elaborated in detail in Chap. 2) is facilitated by the use of a common mathematical formalism. Notably, as Ehlers in Ehlers (1995, p. 175) remarked, it removes "the 'incommensurability' of concepts of different theories, in particular, Newton's and Einstein's theories of spacetime and gravitation". A common mathematical framework makes it especially clear in which way the mathematical structures on the different levels of spacetimes are degenerate special structures of each other. On the physical side, the physical interpretation of different aspects of Newtonian-like and relativistic theories is now possible in terms common to all physical theories under investigation. Concerning the law of inertia, as was pointed out in Earman and Friedman (1973, p. 336):

> From the four-dimensional point of view it is hardly surprising that the foundations of the first law are intimately connected with issues about the structure of spacetime.

As a result of such a four-dimensional geometrical analysis, the prevalent Newtonian theories emerge, in a sense, "to be as four-dimensional" (Friedman 1983) as special and general relativity, and a covariant formulation—which is possible for any physical theory, as recognized first by E. Kretschmann in 1917 in Kretschmann (1917)—of Newtonian mechanics is as 'natural' as the usual 3-vector calculus presentation. [Einstein was quite sceptical about the possibility of a fully covariant and 'simple' formulation of Newtonian mechanics (Einstein 1918).] Conceptually, Einstein's theory of gravitation is much simpler than almost all non-relativistic spacetime theories, mainly because there are many more geometrical objects defined on the associated spacetime manifold, with specific laws and field equations governing them. To summarize the four-dimensional point of view, we quote from R. Penrose's profound analysis of the structure of spacetime in Penrose (1967, p. 7):

> However there can be little doubt that of all views of space and time that have been put forward up to the present day, the Einsteinian view is the most comprehensive, the most profound, and also the most accurate.

In this and the following section, we present three classical, non-relativistic types of spacetime theories and their associated geometrical structures: Leibnizian, Galilean, and Newtonian spacetime [in the hypothetical absence of gravitation, i.e., in empty (flat) space, and in the presence of gravitational fields]. Surprisingly, it is possible to give a formulation of Newtonian theory in which gravitational forces are 'geometrized away' in the spirit of general relativity, i.e., the universal force of gravitation is 'absorbed' into the (curved) geometry. Mathematically, this amounts to incorporating gravity in a now non-flat affine connection of spacetime. With such a reformulation of Newtonian gravity to hand (of course, possible only after Minkowski's unification of space and time in 1908), Einstein's path towards a general-relativistic theory of gravity would presumably have required a less giant step to his final version of general relativity. In the words of Penrose in Penrose (1968, p. 143) and Penrose (1967, p. 15), respectively:

> On the other hand, if Newtonian theory had been reformulated as a *space-time theory* [...], then a quite different alteration—namely, to some form of general relativity—would have appeared as mathematically natural.

> If Einstein had never existed but Newton's gravitational theory had been reformulated as a space-time theory (a very big "but"!), then the discovering of general relativity would not seem so very great a step, perhaps not even requiring Einstein's genius for its achievement!

Furthermore, and along the lines of Klein's Erlangen program of 1872 as established in Klein (1872), we discuss the symmetry groups of transformations or automorphisms (namely the kinematical, the Newtonian, the Galilean, and the elementary group, which are—in reverse order—successive subgroups), preserving the various, more and more restricted, geometrical structures on the spacetimes mentioned above. Where necessary and illuminating, we also give results in the usual 3-vector formalism.

Common to Leibnizian, Galilean, and Newtonian theory is the notion of a *spacetime* or Minkowski's 'world' $\mathcal{M}$, the union of all possible physical 'events' or worldpoints. So, to begin with, $\mathcal{M}$ is actually a *set*. Mathematically, the spacetime is in fact represented as a four-dimensional, real, connected and differentiable, affine *manifold* $\mathcal{M}$. The manifold picture enables one to coordinatize spacetime by a quadruple of four real numbers (one for time and three for each point in space), and also to provide coordinate systems $\langle x^\mu \rangle$. The basic structure on $\mathcal{M}$ is that of a *topology*, i.e., roughly speaking a statement about the neighborhood of each point $p \in \mathcal{M}$, and open sets $\mathcal{U} \subset \mathcal{M}$. Given the standard (locally Euclidean) topology of a manifold, following Friedman (1983), we formulate all subsequent structures and laws as purely local statements, i.e., as field equations, leaving aside (quite interesting and often difficult) questions about the global aspects and properties of spacetime.

Furthermore, the manifold $\mathcal{M}$ is supposed to have an *affine structure* or (a linear and symmetric) *connection*. This affine connection provides the notion of an infinitesimal parallel displacement of directions at each point in $\mathcal{M}$, i.e., it is possible to relate vector spaces (the tangent spaces $T_p\mathcal{M}$) of 'nearby' points. The term 'affine manifold' was introduced by H. Weyl (1919, Sect. 15), who emphasized the independence of the connection from any given metric, e.g., the whole tensor calculus on manifolds simply rests on the basic idea of an affine notion of infinitesimal parallel displacement (see also Weyl 1923, p. 17). Mathematically, the connection may be introduced by a derivative operator ∇_μ, a generalization of the usual directional derivative of a vector in a Euclidean vector space (see, e.g., Friedman 1983, App. 5). For every tangent vector v^μ of a curve $\sigma(\tau)$ in $\mathscr{M}$ with parameter τ, we can measure the changes of direction of the curve itself, namely the acceleration field,

$$a^\mu = v^\kappa \nabla_\kappa v^\mu = \frac{\mathrm{d}v^\mu}{\mathrm{d}\tau} + \Gamma^\mu_{\nu\lambda} v^\nu v^\lambda = \frac{\mathrm{d}^2 x^\mu}{\mathrm{d}\tau^2} + \Gamma^\mu_{\nu\lambda} \frac{\mathrm{d}x^\nu}{\mathrm{d}\tau} \frac{\mathrm{d}x^\lambda}{\mathrm{d}\tau} , \tag{1.5}$$

with respect to an arbitrary coordinate system $\langle x^\mu \rangle$, and with $\Gamma^\mu_{\nu\lambda}$ being the components of the connection ∇_μ relative to this coordinate system, defined by

$$\nabla_{\partial/\partial x^\nu} \frac{\partial}{\partial x^\lambda} = \Gamma^\mu_{\nu\lambda} \frac{\partial}{\partial x^\mu} .$$

(A *curve* in spacetime is a one-dimensional submanifold with a parameter associated with it, whereas such a distinguished parameter does not exist on a *path*.) Curves in spacetime satisfying $a^\mu = v^\kappa \nabla_\kappa v^\mu = 0$ are called *geodesics*. These are also *autoparallels*. Geometrically, every tangent direction of the curve is infinitesimally parallel transported, and the tangent vectors of the curve are constant along the curve. In an affine manifold, these 'straight' worldlines or 'affine straights' constitute a privileged class of paths. The affine perspective is put in a nutshell by A. Trautman in Trautman (1965, p. 103):

> There is another fundamental principle of physics that is common to all physical theories so far put forward. This is that the differentiable manifold of space and time is endowed with an affine connection whose geodesics form a privileged set of worldlines in spacetime. The particular affine connection depends on the theory we are considering, but the existence of an affine connection is common to all theories. It is necessary in order that the fundamental law of physics can be expressed in the form of differential equations, which is certainly true of all physical theories.

However, the affine structure does not provide a notion of length and angle. So far, neither of these have meaning on $\mathscr{M}$.

In order to link space and time in a way specific to the different 'classical' physical theories of Newton, Leibniz, and Galileo, this primitive concept of the actual world has to be endowed with additional, intrinsic geometrical structures. [For a critical account of this process of structuring spacetime, initially considered as a set, according to various physical inputs, see Giulini (2009); for details of the different levels of structure in general relativity, see Chap. 2.] We start with 'empty' *Newtonian spacetime*: space is either completely devoid of any matter or else matter is assumed to be so sparsely distributed in space that we can, at least to a good approximation, neglect gravity. In this way, Newtonian spacetime has the character of a mere rigid background spacetime, in which gravity plays no dynamical role. In mathematical terms, our manifold $\mathscr{M}$ has to be flat or Euclidean, i.e., the affine manifold $\mathscr{M}$ is in fact globally diffeomorphic to the four-dimensional *affine space* $\mathbb{E}^4$.

Associated with the affine connection ∇_μ is the Riemann–Christoffel curvature tensor, with components

$$R^\mu_{\ \nu\lambda\kappa} = \frac{\partial}{\partial x^\lambda}\Gamma^\mu_{\nu\kappa} - \frac{\partial}{\partial x^\kappa}\Gamma^\mu_{\nu\lambda} + \Gamma^\sigma_{\nu\kappa}\Gamma^\mu_{\sigma\lambda} - \Gamma^\sigma_{\nu\lambda}\Gamma^\mu_{\sigma\kappa}\,,$$

in an arbitrary coordinate system $\langle x^\mu \rangle$ (see also Sects. 2.2 and 2.4 for the definition of the Riemann tensor in relativistic spacetimes). There is a coordinate system around $p \in \mathcal{M}$ in which the components of the affine connection vanish, if and only if the curvature tensor vanishes at that point (see, e.g., Laugwitz 1965, p. 109f). In this case, the geodesic law (1.5) simplifies to

$$a^\mu = v^\kappa \nabla_\kappa v^\mu = \frac{\mathrm{d}^2 x^\mu}{\mathrm{d}\tau^2} = 0\,, \tag{1.6}$$

whence $x^\mu(\tau) = v^\mu(0)\tau + x^\mu(0)$, with constants $v^\mu(0)$ and $x^\mu(0)$. These solutions of (1.6) are just the familiar Euclidean straight lines. The equation of motion (1.6) is Newton's law of inertia, and the coordinate system $\langle x^\mu \rangle$ with $\Gamma^\mu_{\nu\lambda} = 0$ is called an *inertial system*. Recall once again, that Newton's first law does not in fact provide a privileged parametrisation of spacetime paths, since the structure it bestows upon spacetime is merely projective. This observation can probably be attributed to H. Weyl in Weyl (1919) (on p. 324 of the eighth edition of 1993):

> If one deprives the metric of its original character, it is my opinion that experience gives no further clue about how to ascribe an affine connection to the world. Inertial motion merely indicates a 'projective nature', according to which there exists a process of infinitesimal parallel displacement of *directions in themselves.*

In the affine point of view, the worldlines of free particles ('free motions') are just given by the geodesics (timelike straight lines) of the affine connection. As Ehlers put it so clearly in Ehlers (1973a, p. 75):

> This axiom is a precise formulation of the law of inertia, which emphasizes its intrinsic, coordinate-independent content. [... Newton's] law of inertia serves to define the affine structure of spacetime. The subsequent laws of dynamics presuppose that structure but do not restrict or enrich the spacetime geometry further.

[See also Straumann (1987, p. 31f) and Trautman (1965, p. 103f), who, with Ehlers, both emphasize that Newton's law of inertia is "not a trivial consequence of Newton's second law", but "is in fact one of the most important laws of physics."]

On $\mathcal{M}$, we have two other distinguished tensor fields, which define the metric structure of $\mathcal{M}$, namely a temporal and a spatial 'metric'. First, there exists—unique up to linear transformations—an *absolute time*, a continuous linear map $t : \mathcal{M} \to \mathbb{R}$ with non-vanishing gradient. This Newtonian time function t—mathematically representing the readings of standard or 'world' clocks assigned to an event of the world—defines the simultaneity relation $t(p) = t(q)$, and a time-order $t(p) \gtrless t(q)$ between two events $p, q \in \mathcal{M}$, and furthermore the durations of processes or the temporal distance $|t(p) - t(q)|$ between any two worldpoints. Associated with the coordinate function $x^0 = t$, we have a temporal covector field $\mathrm{d}t : T_p\mathcal{M} \to \mathbb{R}$ with components given by the gradient of the scalar field $t_\mu = \partial t/\partial x^\mu$, and therefrom a symmetric, but singular *temporal metric* $dt \otimes dt : T_p\mathcal{M} \times T_p\mathcal{M} \to \mathbb{R}$ of type $(0,2)$, with components

$$t_{\alpha\beta} = t_\alpha t_\beta = \frac{\partial t}{\partial x^\alpha} \frac{\partial t}{\partial x^\beta} .$$

Having fixed the physical unit in which time is measured, the covector field dt uniquely represents the above-mentioned family of time-functions t, which are fixed up to the choice of origin only. Every tangent vector $X \in T_p\mathcal{M}$ is called *future-* or *past-directed timelike* if $dt_p(X) \gtrless 0$, and *spacelike* if $dt_p(X) = 0$. The latter are the standard three-vectors $\mathbf{x} \in \mathbb{R}^3 \simeq T_p\mathbb{R}^3$ in the usual textbook formulations of Newtonian theory. The tensor field $dt \otimes dt$ assignes a nonzero value

$$|X| = \sqrt{dt(X) \otimes dt(X)} = \sqrt{t_\alpha t_\beta x^\alpha x^\beta}$$

only for timelike vectors $X = (x^0, x^1, x^2, x^3) = (x^0, \mathbf{x})$. The temporal metric then induces a temporal duration or time difference between any two points $p, q \in \mathcal{M}$ in different subsets (hyperplanes) of simultaneous events

$$\mathscr{S}_p = \{e \in \mathcal{M} \mid t(e) = t(p) = \tau\} ,$$

given by the above simultanity relation. Spacetime is stratified in successive three-dimensional spacelike sections $\mathscr{S}_p = \{X \in T_p\mathcal{M} \mid dt_p(X) = 0\}$, these being the level surfaces of the gradient of the scalar field t. The temporal metric also determines the *causal structure*, due to the fact that the future and past of any event e have a common boundary, namely the hyperplane $\mathscr{S}_e$ representing the present. Furthermore, we require the covariant derivative of the covector field dt to vanish with respect to the affine connection, i.e., compatibility with the given affine structure, which in coordinates reads $t_{\alpha\beta;\gamma} = 0$ and $t_{\alpha;\gamma} = 0$. In an inertial coordinate system $\langle x^\mu \rangle$, the four-velocity of any timelike curve is well-defined, with

$$(v^\mu) = (1, \mathbf{v}) = (1, dx^1/dt, dx^2/dt, dx^3/dt) .$$

The acceleration four-vector is (always!) spacelike and has components

$$(a^\mu) = (0, \mathbf{a}) = (0, d^2x^1/dt^2, d^2x^2/dt^2, d^2x^3/dt^2) ,$$

with respect to an inertial system.

Secondly, to have a notion of length and angle, on each $\mathscr{S}_p$ of the above defined foliation there exists a positive definite *spatial metric* dl_p^2 which ensures that each spacelike section $\mathscr{S}_p$—*space*—is indeed a Euclidean 3-space. Mathematically, dl_p^2 represents length measurements by standard measuring rods. These 'inner' or scalar products between spacelike vectors are identical on each section. The components of dl_p^2 with respect to a Cartesian reference frame are given by the familiar Kronecker symbol δ_k^i. Physically, the Euclidean geometry of space is maintained by passing from one instant of time to another. Associated with the three-dimensional metrics dl_p^2 at each 'instant' $\mathscr{S}_p$, there is a symmetric, but also singular tensor field ds^2 : $T_p^*\mathscr{M} \times T_p^*\mathscr{M} \to \mathbb{R}$ of type $(2,0)$ on the whole manifold $\mathscr{M}$, with components $s^{\alpha\beta}$ relative to a coordinate system $\langle x^\mu \rangle$. [See, e.g., Friedman (1983, p. 76ff) for details of the construction of this tensor.]

As noted by Malament (1986a, Sect. 2), up to now, the affine structure or connection only gives a notion of 'constancy' of vectors, e.g., of the acceleration four-vector of a timelike curve defined by (1.6), but no information about the 'magnitude' of this vector. The spatial metric $s^{\alpha\beta}$ serves to assign such a length to spacelike vectors only, but not to timelike ones! In particular, inserting the covector dt into the singular metric ds^2 results in $s^{\alpha\beta}t_\alpha t_\beta = 0$, which is equivalent to the 'orthogonality' relation $s^{\alpha\beta}t_\alpha = 0$. As for the temporal metric, we require the spatial metric to be compatible with the flat connection, $s^{\alpha\beta}{}_{;\gamma} = 0$. Now, the 'metrical' character of the spacetime $\mathscr{M}$ is fixed by the pair of the two singular metric tensors $(dt \otimes dt, ds^2)$, which is called a *Newtonian metric*. In a Cartesian coordinate system, the metric structure on $\mathscr{M}$ is given by $(t_\alpha) = (\delta_\alpha^0) = (1,0,0,0)$ and the matrix $(s^{\alpha\beta}) = \mathrm{diag}(0,1,1,1)$.

In summary, a collection $(\mathscr{M}, \nabla_\mu, t_{\alpha\beta}, s^{\alpha\beta})$, satisfying the 'compatibility' and 'orthogonality' conditions stated above is called a *classical spacetime* structure. Physically, the compatibility conditions state agreement upon the notion of 'constancy' induced by either the metric or the affine structure, both coexisting on the manifold $\mathscr{M}$. In detail, spatial compatibility $\nabla_\gamma s^{\alpha\beta} = 0$ ensures that, if some 'infinitesimal' measuring rod (represented by a spacelike vector connecting two simultaneous events) does not experience any motion relative to an observer (represented by a timelike curve)—the state of motion being determined by ∇_μ—then the ruler also experiences no change in length (as measured by $s^{\alpha\beta}$). Equivalently, temporal compatibility $\nabla_\gamma t_{\alpha\beta} = 0$ means that the 'temporal length' (as measured by $t_{\alpha\beta}$) of a constant vector field—again determined by ∇_μ—along any timelike curve is constant.

To define a standard of absolute rest and of no rotation—a 'rigging' of Newtonian spacetime—a further geometrical object is needed, viz., a distinguished, but essentially arbitrary timelike vector field t, which we normalize by $dt(t) = t_\mu t^\mu = 1$ in order that two points e, $p = e + tt$ are separated by the time span t, i.e., Newtonian clocks exactly measure the coordinate time t. Compatibility of t with the connection, i.e., $t^\alpha{}_{;\gamma} = 0$, states that the timelike lines of our rigging are in fact geodesics. On the one hand, the vector field t defines a time axis, called 'absolute time' $\mathscr{T}$. On the other hand, it fixes events e in different space sections $\mathscr{S}_e \neq \mathscr{S}_p$ to be at the same point in space at different times by the host of all timelike lines $\{p \in \mathscr{M} \mid p = e + tt, t \in \mathbb{R}\}$. This family of non-intersecting geodesics, slicing the planes of simultaneity, thereby define 'absolute rest' [see Fig. 1.3, adapted from Friedman (1983)]. From all "relative spaces"—in the words of Newton—this (arbitrary!) vector field t essentially singles out one to be 'absolute space' $\mathscr{S}$.

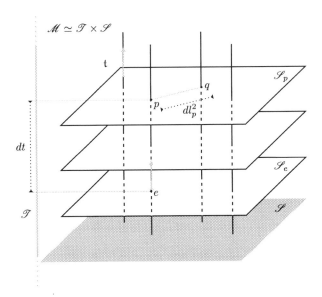

Fig. 1.3 Rigging of Newtonian spacetime $\mathcal{M}$ in $2 + 1$ dimensions by a timelike vector field t, defining a family of timelike geodesics: for any two points $e, q \in \mathcal{M}$ in different sections $\mathscr{S}_e$ and $\mathscr{S}_p$, the distance between these two points is just the distance between points p and q within the same spacelike section $\mathscr{S}_p$, with p located at the same space point as e according to the rigging. The distance between e and q is then measured by the three-dimensional metric dl_p^2 in $\mathscr{S}_p$. However, this construction is 'singular', because the distance between, e.g., the different spacetime points p and e vanishes

According to (dt, t), for every vector $X \in T_p\mathcal{M}$ we now have a unique decomposition: $X = dt_p(X)\mathsf{t} + \mathbf{x}$ with $dt_p(\mathbf{x}) = 0$ (see, e.g., Kriele 1999). Therefore, the projection $\mathbf{x}$ of X onto the 3-vector space $\mathscr{S}_p$ is spacelike. A four-vector X with $|\mathbf{x}| = 0$ is called *null*. A particle, represented by a timelike worldline, with null tangent vectors is therefore at 'absolute rest', i.e., all points on the trajectory are at the same space point. In particular, as emphasized by Friedman (1983, p. 74f),

> The rigging is not chosen to be 'orthogonal' to the planes of absolute simultaneity; rather the choice of rigging itself defines such 'orthogonality'.

Collecting all our local field equations of Newtonian spacetime, we end up with a system of six equations for four geometrical objects (tensor fields), namely a linear, symmetric connection $\Gamma^\mu_{\nu\lambda}$ and associated derivative operator ∇_μ, a temporal and a spatial (Newtonian) metric $(t_{\alpha\beta}, s^{\alpha\beta})$, and one vector field t^μ, defined on the entire manifold $\mathcal{M}$:

$$R^\alpha{}_{\beta\gamma\delta} = 0 \,, \tag{1.7}$$

$$t_{\alpha;\gamma} = 0 \,, \tag{1.8}$$

$$s^{\alpha\beta}{}_{;\gamma} = 0 \,, \tag{1.9}$$

$$s^{\alpha\beta} t_\alpha = 0 \, , \tag{1.10}$$

$$t^\alpha{}_{;\gamma} = 0 \, , \tag{1.11}$$

$$t_\mu t^\mu = 1 \, . \tag{1.12}$$

According to (1.7), the four-dimensional spacetime $\mathscr{M}$ is essentially flat. Equations (1.8), (1.9), and (1.11) form a group of 'connection axioms', in particular expressing compatibility of the affine and the (spatial and temporal) metrical structure. The 'orthogonality' relation between the temporal covector field and the spatial metric is expressed by (1.10). The remaining (1.12) is a kind of 'normalization' of the rigging of spacetime.

Summarizing the assumptions made above, Newtonian spacetime—as a four-dimensional manifold—is structured geometrically in the following way: $\mathscr{M}$ is stratified into 'horizontal' sequences of Euclidean hyperplanes of absolute simultaneity $\mathscr{S}$ (three-dimensional 'absolute space' or 'absolute rest') and fibred 'vertical' by the rigging $\mathscr{T}$ (one-dimensional 'absolute time') [see Fig. 1.4, adapted from Ehlers (1973a)]. Mathematically, the flat manifold $\mathscr{M}$, diffeomorphic to the

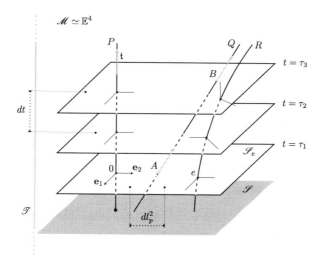

Fig. 1.4 Newtonian spacetime $\mathscr{M}$ in $2 + 1$ dimensions—one space dimension being suppressed: stratification of affine space $\mathbb{E}^4$ into absolute space $\mathscr{S} \simeq \mathbb{E}^3$ and absolute time $\mathscr{T} \simeq \mathbb{E}^1$. Illustrated are three successive planes of absolute simultaneity at different times $t = \tau_1 < \tau_2 < \tau_3$. On each spacelike section, distance measurements are possible due to a spatial Euclidean metric dl_p^2. The time difference between any two different 3-spaces is measured by a Newtonian clock, the covector field dt. Vector parallelism of 4-vectors A and B is provided by Newton's first law, which is encoded in a flat affine connection on $\mathscr{M}$. The worldline of observer P is parallel to the timelike vector field t, defining absolute rest, with the tangent vectors of P being null. The axes e_1, e_2 of a Cartesian coordinate system transported along this curve are non-rotating (Newtonian frame), representing an inertial system. The worldline of Q is a straight line, Q moving with constant velocity (with the tangent vectors of Q non-null), whereas observer R is accelerating

affine space $\mathbb{E}^4$, has a product structure $\mathcal{M} \simeq \mathbb{E}^1 \times \mathbb{E}^3 \simeq \mathcal{T} \times \mathcal{S}$, the Cartesian product of all instants of time $\mathcal{T}$ and all space points $\mathcal{S}$, with $\mathbb{E}^1$ a one-dimensional affine space and $\mathbb{E}^3$ an affine 3-space (in fact these affine spaces are Euclidean spaces). In essence, Newtonian spacetime is a quadruple $(\mathbb{E}^4, t, dt, ds^2)$ satisfying the local field equations (1.7)–(1.12). From the four-dimensional point of view, Newtonian theory, although ostensibly quite intuitive from a physical standpoint, turns out to be a rather complicated mathematical concept!

Next, we consider the spacetime structure of classical, non-relativistic kinematics, referred to as *Leibnizian spacetime* by Ehlers in Ehlers (1973a, 1995). In this world model—opposed to Newton's theory—the existence of a standard of absolute rest and no rotation is negated on the basis of the invariance of physical phenomena with respect to velocity transformations, as already mentioned in Sect. 1.1, i.e., the relativity of motion. Mathematically, there is no notion of an infinitesimal parallel displacement along timelike paths on $\mathcal{M}$, i.e., no vector parallelism of 4-vectors. This lack of an affine structure in Leibnizian spacetime has the consequence that physically there are no preferred motions (no straight lines) or inertial systems (non-rotating reference systems) [see Fig. 1.5, adapted from Ehlers (1973a)]. So, in fact, $\mathcal{M}$ is not an affine manifold. As in the Newtonian case, $\mathcal{M}$ is flat, and we have a metrical structure on $\mathcal{M}$: again an absolute time function t and corresponding covector field dt, and an associated singular temporal metric $dt \otimes dt$. Further, in

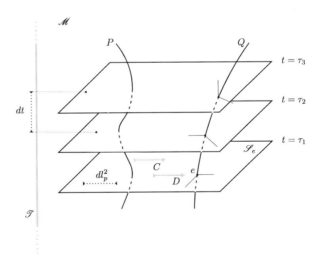

Fig. 1.5 Leibnizian spacetime $\mathcal{M}$ in $2+1$ dimensions—one space dimension being suppressed: illustrated are three successive planes of absolute simultaneity at different times $t = \tau_1 < \tau_2 < \tau_3$. On each spacelike section, distance measurements are possible due to a spatial Euclidean metric dl_p^2. The time difference between any two different 3-spaces is measured by a world clock, the covector field dt. Vector-parallelism of 4-vectors is not defined, due to the lack of an affine connection on $\mathcal{M}$. In contrast, there is standard $\mathbb{R}^3$-parallel transport of 3-vectors C and D in each hyperplane. The relative velocity $\mathbf{v}$ or acceleration $\mathbf{a}$ between two observers P and Q are spacelike vectors and hence well-defined

all the spacelike sections S_p of absolute simultaneity there exist identical spatial, Euclidean metrics dl_p^2, which are again merged into a single, singular tensor field ds^2 on the whole manifold $\mathcal{M}$, with $s^{\alpha\beta}t_\alpha = 0$ to mesh with the temporal covector field dt.

Compared to the previous model of the world, Leibnizian spacetime is a rather straightforward setting. As Ehlers remarks aptly in Ehlers (1973a, p. 75):

> Leibniz's spacetime, the spacetime of nonrelativistic kinematics, has less structure than Newton's. [...] Leibniz's kinematical spacetime has not enough, Newton's dynamical spacetime has too much structure.

In essence, this structure is well suited from a relational point of view—in this way realizing Leibniz's relative conception of space—since it is only meaningful to make relative statements, e.g., about the relative velocity, acceleration, angular velocities, etc., between two observers. [See Earman (1989) for an overview of different concepts of classical spacetimes, their structures, invariants, and questions which it is meaningful to ask in these theories only.]

Lange's work on the law of inertia, as outlined in Sect. 1.2, provided the profound insight that absolute space $\mathcal{S}$ and absolute rest are in fact superfluous elements of Newton's theory, which are not presupposed by nature. The world model which rests upon this change in structure, retaining the stratification of spacetime into hypersurfaces of constant 'absolute time' (planes of absolute simultaneity), is *Galilean spacetime*, the spacetime of classical dynamics. Mathematically, Galilean spacetime is readily obtained from Newton's spacetime by omitting the timelike vector field t which defines 'absolute rest' and the relation of 'occurring-at-the-same-place'. As a first consequence, the notion of 'rest' is only meaningful relative to a single inertial observer. A particle at rest relative to a particular inertial reference system is generally moving uniformly with respect to another inertial system. Secondly, for any two points $p, q \in \mathcal{M}$ in different hyperplanes, the distance between these points is meaningless because "only simultaneous events have an unambiguous spatial distance, in consequence of the 'relativity of space', i.e., the absence of any objective criterion for two nonsimultaneous events to happen at 'the same' space point" (Ehlers 1973b, p. 4).

As in Newtonian theory, space in our Galilean world has a Euclidean geometry and the affine manifold $\mathcal{M}$ is flat. The field equations governing Galilean spacetime boil down to (1.7)–(1.10) from Newtonian spacetime, with all the structures (except the rigging) already known from Newtonian theory. In essence, a collection $(\mathbb{E}^4, dt, ds^2)$, is called a *Galilean spacetime*. The metrical character of $\mathcal{M}$ is again fixed by a pair of singular metric tensors $(t_{\alpha\beta}, s^{\alpha\beta})$, which is called a *Galilean metric* on $\mathcal{M}$. Still, the worldlines of privileged objects, 'free particles', correspond to privileged paths, 'free motions' [see Fig. 1.6, adapted from Ehlers (1973a)]. These motions are determined by the affine structure via the geodesic law of (1.5) which, in an inertial system reduces to (1.6): free particles move uniformly on a straight line!

Finally, to conclude the analysis of classical, non-relativistic spacetimes, we classify the groups of transformations preserving their different geometrical structures,

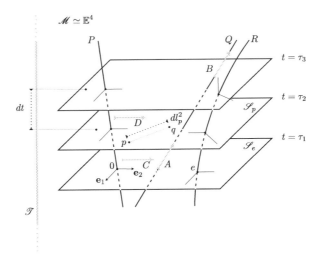

Fig. 1.6 Galilean spacetime $\mathscr{M}$ in $2 + 1$ dimensions—one space dimension being suppressed: illustrated are three successive planes of simultaneity at different times $t = \tau_1 < \tau_2 < \tau_3$. The lack of a distinguished timelike vector field t forces one to abandon absolute space $\mathscr{S}$ and absolute 'rest'. A statement about being at 'rest' is only meaningful relative to a single inertial observer. The distance measured by dl_p^2 in $\mathscr{S}_p$ is meaningful only for two points p and q in this plane of absolute simultaneity. Parallelism of timelike 4-vectors A and B and of spacelike 4-vectors C and D is established. The worldlines of particles P and Q in uniform motion are straight (defined by the affine structure and Newton's first law), whereas observer R is accelerating

a group-theoretical approach in the spirit of Klein's programme mentioned in the introduction to this section (see Ehlers 1973a,b and Earman 1989, Chap. 2).

To begin with, we consider Newtonian spacetime (in the absence of gravity), i.e., $(\mathbb{E}^4, t, dt, ds^2)$ with its global Cartesian product structure $\mathscr{T} \times \mathscr{S}$ of time and space. The group of automorphisms is the direct product of dilations, rotations, and translations of $\mathscr{S}$ with the affine group of $\mathscr{T}$, referred to by Weyl (1919, Sect. 20) as the 'elementary group' $\mathfrak{E}$ in space and time. Mathematically, this is a 9-parameter Lie group: two inertial coordinate systems $\langle x^\mu \rangle$ and $\langle x'^\mu \rangle$, used to label events, are related to each other by the mappings $x^j \to x'^j = R^j{}_k x^k + c^j$ (a rotation followed by a translation in space), $x^j \to x'^j = dx^j$ (a dilation in space), and $t \to at' + b$ (a dilation followed by a translation in time), where $R^j{}_k$ is a time-independent orthogonal matrix, and $a > 0$, b, c, and $d > 0$ are constants. [Disregarding space and time dilations, the symmetry group of Newtonian spacetime is a further restricted, 7-parameter Lie group, the semidirect product of the symmetry group of rigid motions in space, the Euclidean group, and the time translation group (Friedman 1983, p. 84ff).]

The elementary group $\mathfrak{E}$ is a subgroup of the Lie group $\mathfrak{G}$, the 10-parameter invariance group of Galilean spacetime or 'Galilean group', and it is also the invariance group of the laws of classical mechanics according to Galileo's principle of relativity (see Sect. 1.1). Due to the relativity of motion, 'Galilean frames' transform according to the well-known Galilean transformations, $x^j \rightarrow x'^j = R^j{}_k x^k + v^j t + c^j$ (a rotation followed by a velocity transformation and a translation in space), and $t \rightarrow t' + b$ (a shift in time) with $R^j{}_k$, b, and c^j as above and v^j being further constants.

The most general symmetry transformations within our (flat) non-relativistic spacetimes comprise the 'kinematical group' $\mathfrak{K}$, the invariance group of Leibnizian spacetime. This group is not a Lie group since, besides three parameters, its elements, the transformations $x^j \rightarrow x'^j = R^j{}_k(t) x^k + c^j(t)$ and $t \rightarrow t' + b$ relating the coordinates of two reference frames, depend on six real *functions* of time, namely a time-dependent angular velocity and a time-dependent translation velocity.

In summary, the three groups of spacetime automorphisms are successive subgroups with $\mathfrak{E} \subset \mathfrak{G} \subset \mathfrak{K}$. As observed by Earman (1989, p. 36), "as the spacetime structure becomes richer, the symmetries become narrower", or in the words of Friedman (1983, p.64), "the larger the symmetry group, the fewer the absolute elements postulated by a given [spacetime] theory".

In the next section we analyze Newtonian spacetime in the presence of gravity (Newton–Cartan or 'general non-relativistic' spacetime). This is described essentially by a non-flat connection. As we will see, in such a curved spacetime model, there are no longer any global inertial frames, only local ones. The automorphisms comprise time-independent rotations, but time-dependent translations, $x^j \rightarrow x'^j = R^j{}_k x^k + c^j(t)$ and $t \rightarrow t' + b$, these being elements of the 'Newtonian group' $\mathfrak{N}$. The corresponding (non-rotating) coordinate systems are called 'Newtonian frames'. We thus get the hierarchical succession $\mathfrak{E} \subset \mathfrak{G} \subset \mathfrak{N} \subset \mathfrak{K}$ characterizing our different spacetime geometries. In the words of Ehlers (1973a, p. 79), this constitutes

[...] a relation which indicates in a condensed form the evolution of spacetime concepts at the nonrelativistic level. Whereas the transition from $\mathfrak{E}$ to $\mathfrak{G}$ represents the preliminary compromise between the absolutist Newton ($\mathfrak{E}$) and the relativist Leibniz ($\mathfrak{K}$), the step from $\mathfrak{G}$ to $\mathfrak{N}$—or from a flat to a curved connection—is a (somewhat delayed) response to Mach's criticism of the unfounded distinction between inertia and gravity.

1.6 Newton's Law of Gravitation, and Poisson's Equation

In this section gravity enters the scene by transition from hitherto empty space to a 'matter filled' world. First, we reformulate Newton's law of gravitation in its local differential form, namely Poisson's equation, in a four-dimensional, geometric setting, as first undertaken in Cartan (1923) and Friedrichs (1928). We shall also make some brief remarks on Newton's second law. Next, we illuminate the characteristic and special features of Newton's inverse-square law of gravitation, and its observational and experimental evidence. Finally, we outline some still open problems in connection with static and rotating stars in Newtonian gravity, the latter probably being the most common visible objects in our universe.

To begin, we try to incorporate gravity in a satisfactory way into the spacetime geometry. Conceptually, we proceed by passing from *kinematics* to *dynamics*. As geometric structures for Newtonian theory in the presence of gravity we choose, following Ehlers (1973a) and Friedman (1983, Sects. III.3 and III.4), from our three nonrelativistic spacetime models of the previous section those of Galilean spacetime. [For a very recent, concise mathematical textbook presentation of the 'geometrized' formulation of Newtonian gravitation theory, see also Malament (2012, Chap. 4).]

A particle with inertial mass m_i and four-velocity v^μ, subjected to a force field F^μ, obeys the law of motion

$$m_i a^\mu = m_i(v^\kappa \nabla_\kappa v^\mu) = m_i \left(\frac{\mathrm{d}^2 x^\mu}{\mathrm{d}t^2} + \Gamma^\mu_{\nu\lambda} \frac{\mathrm{d}x^\nu}{\mathrm{d}t} \frac{\mathrm{d}x^\lambda}{\mathrm{d}t} \right) = F^\mu , \qquad (1.13)$$

where F^μ is spacelike. In an inertial system $(F^\mu) = (0, \mathbf{F})$, and (1.13) is just Newton's second law $\mathbf{F} = m_i \mathbf{a}$ in standard 3-vector notation. [The proportionality of force and acceleration in Newton's second law has been tested in the limit of small forces and accelerations by Gundlach et al. (2007) down to accelerations of 5×10^{-14} m/s^2.] If the particle with passive gravitational mass m_g moves in a gravitational field with gravitational field strength $\mathbf{g} = -\nabla\Phi$ and associated gravitational potential Φ, we have in the spacelike sections of absolute simultaneity, and with respect to a non-rotating frame of reference, $m_i \ddot{\mathbf{r}} = -m_g \nabla\Phi + \mathbf{F}$, where $\mathbf{F}$ subsumes all non-gravitational force fields. The dynamical field Φ is governed by an elliptic differential equation, namely Poisson's equation $\Delta\Phi = 4\pi\varrho$, where the scalar field ϱ is the mass density and the single source of the gravitational field in Newtonian physics. (In particular, and in contrast to general relativity, there are no contributions coming from any momentum, or even stresses.) In empty space, Poisson's equation reduces to the familiar Laplace equation $\Delta\Phi = 0$, whose solutions Φ are the so-called harmonic functions.

It is a standard textbook calculation to transform Newton's second law $\mathbf{F} = m_i \mathbf{a}$ from an inertial frame into a translationally and rotationally accelerated, noninertial reference frame (see, e.g., Marsden and Ratiu 1999, Sect. 8.6), with the result

$$\ddot{\mathbf{r}} = \frac{m_g}{m_i} \mathbf{g} - \frac{1}{m_i} \mathbf{F} - \mathbf{a} - \dot{\boldsymbol{\omega}} \times \mathbf{r} - 2\boldsymbol{\omega} \times \dot{\mathbf{r}} - \boldsymbol{\omega} \times (\boldsymbol{\omega} \times \mathbf{r}) , \qquad (1.14)$$

where $\mathbf{a}$ and $\boldsymbol{\omega}$ are the translational and rotational accelerations, respectively, relative to the inertial frame. The last three terms on the right-hand side of (1.14) are the well-known *Euler, Coriolis,* and *centrifugal accelerations*. As emphasized in Ehlers (1973b), if nature reveals that the ratio m_g/m_i is a universal constant, it is impossible to measure separately the gravitational field strength $\mathbf{g}$ and the translational acceleration $\mathbf{a}$, but only the combination $\mathbf{g}-\mathbf{a}$. Therefore, already at this level, it is quite clear, that a division into pure gravitational and pure translational acceleration fields is unjustified on empirical grounds.

Now, in the four-dimensional formulation on the fixed Galilean spacetime, we get a system of five equations for five geometrical objects, namely a flat, linear, symmetric affine connection $\Gamma^{\mu}_{\nu\lambda}$, a pair of singular, symmetric (temporal and spatial) metrics $(t_{\alpha\beta}, s^{\alpha\beta})$, and two scalar fields, the gravitational potential Φ and the mass density ϱ (see Earman and Friedman 1973 and Friedman 1983, Sect. III.3). The first four equations govern Galilean spacetime in the (hypothetical) absence of gravity, as specified by (1.7)–(1.10) of the previous section, and these are complemented by the four-dimensional generalization of Poisson's equation, viz.,

$$s^{\alpha\beta}\Phi_{;\alpha;\beta} = 4\pi\varrho .\tag{1.15}$$

Particles solely under the influence of gravity follow spacetime paths obeying

$$m_{\mathrm{i}}\left(\frac{\mathrm{d}^2 x^{\mu}}{\mathrm{d}t^2} + \Gamma^{\mu}_{\nu\lambda}\frac{\mathrm{d}x^{\nu}}{\mathrm{d}t}\frac{\mathrm{d}x^{\lambda}}{\mathrm{d}t}\right) = -m_{\mathrm{g}}\, s^{\mu\kappa}\Phi_{;\kappa} .\tag{1.16}$$

As announced in the introduction to Sect. 1.5, we try to incorporate the gravitational force and the potential into the geometry of spacetime, i.e., to 'geometrize away' gravity in the spirit of general relativity. The key observation is, that the mass cancels from the equation of motion (1.16) obeyed by material particles acted on only by a gravitational force. This is due to the observational fact that the inertial and gravitational mass are equal, i.e., $m_{\mathrm{i}} = m_{\mathrm{g}}$.

Today, this equality of inertial and gravitational mass has been verified by a rotating torsion balance experiment (using beryllium and titanium test bodies) with the exceptional accuracy of $|m_{\mathrm{g}} - m_{\mathrm{i}}|/m_{\mathrm{i}} < 2 \times 10^{-13}$ by Schlamminger et al. (2008). Future free fall tests with satellite based experiments like MICROSCOPE (MICRO-Satellite à traînée Compensée pour l'Observation du Principe d'Equivalence) and STEP (Satellite Test of Equivalence Principle) expect to increase accuracy to 10^{-15} and 10^{-18}, respectively. For an up-to-date textbook presentation of experiments designed to test the equivalence principle, from early measurements using pendulums by Galileo and Newton, through the first torsion balance experiment by Eötvös in 1890, to modern free fall experiments, see Ohanian and Ruffini (2013, Sect. 1.6).

In the spacelike sections of absolute simultaneity, and with respect to suitable, so-called 'non-rotating' or 'Newtonian frames' of reference, free fall motion now reads $\ddot{\mathbf{r}} = -\nabla\Phi$ (see Ehlers 1973b and Friedman 1983, Sect. III.4). In the presence of gravity these 'force-free motions' constitute a preferred set of paths or 'standard motions': All particles with the same initial conditions (same position and velocity) move, irrespective of their masses, in the same way. Newton's second law now may be formulated as

$$m(\ddot{\mathbf{r}} + \nabla\Phi) = \mathbf{F} .\tag{1.17}$$

We quote the central idea of this programme from Ehlers (1973b, p. 5) [see also the related remarks in Ehlers (1973a, p. 77)]:

The transition from kinematics to *dynamics* consists essentially of singling out a particular class of motions as *standard motions* which are considered as force-free, and then to define for all motions differing from standard motions *forces* in terms of accelerations relative to these standard motions. [...] The splitting of the left-hand side of [(1.17)] into an inertial term $m\ddot{\mathbf{r}}$ and a gravitational term $m\nabla\Phi$ is frame-dependent and has no objective significance.

In this way the left-hand side of (1.17) defines (via Newton's second law) the right-hand side, the force function $\mathbf{F}$. The relative acceleration of an arbitrary motion with respect to the standard motions is caused by this force.

In electrodynamics an equivalence between the 'coupling factor' m_g on the right-hand side of (1.16) and the (inertial) mass m_i of the particle, is not true. Charged particles with charge e follow paths obeying

$$m_i\left(\frac{d^2 x^\mu}{dt^2} + \Gamma^\mu_{\nu\lambda}\frac{dx^\nu}{dt}\frac{dx^\lambda}{dt}\right) = e\,F^\mu{}_\nu\frac{dx^\nu}{d\tau}\,,\tag{1.18}$$

where $F^\mu{}_\nu$ is the electromagnetic field tensor. In general, the 'coupling factor' e is very different from the particle's mass m_i, i.e., the ratio e/m_i is not a universal constant. (See, however, the remark in Sect. 1.4 to the effect that the electromagnetic force is, in some respect, the simplest force which is not geometrizable, but which is compatible with the constancy of the light velocity.) This kind of 'non-equivalence' is valid for all other kinds of interactions known in nature, and demonstrates the very distinctive nature of gravity.

Due to the proportionality of inertial and gravitational mass, we can replace, in our four-dimensional spacetime formulation, the flat connection by a nonflat one, whose geodesics are just the trajectories of freely falling particles. These 'free falls' constitute, as mentioned above, a new privileged class of motions (compared to the 'free motions' in the hypothetical absence of gravity, which are the standard motions in the spacetime theories elaborated in Sect. 1.5). Following P. Havas in Havas (1964, 1967), we write this new, curved affine connection $\Gamma^\mu_{\nu\lambda} = \Lambda^\mu_{\nu\lambda} + \Omega^\mu_{\nu\lambda}$. Here $\Lambda^\mu_{\nu\lambda}$ is the previous flat and integrable connection, i.e., $R^\mu{}_{\nu\lambda\kappa}(\Lambda) = 0$, and $\Omega^\mu_{\nu\lambda} = t_\nu t_\lambda s^{\mu\kappa}\Phi_{;\kappa} = t_\nu t_\lambda s^{\mu\kappa}\Phi_{;\kappa}$. In general, and in contrast to general relativity, neither the splitting of this connection into an 'inertial' part $\Lambda^\mu_{\nu\lambda}$ and a 'gravitational' part $\Omega^\mu_{\nu\lambda}$ (determined by the scalar field Φ), nor the flat connection $\Lambda^\mu_{\nu\lambda}$ is unique (unless special boundary conditions are imposed, see below). Any other potential Ψ satisfying $\Psi_{;\mu;[\nu t_\lambda]} = 0$ determines another flat connection $\Gamma^\mu_{\nu\lambda} - s^{\mu\sigma}\Psi_{;\sigma}t_\nu t_\lambda$. Now, with this new connection (altering the affine structure of the spacetime), (1.16) reads

$$a^\mu = v^\kappa\nabla_\kappa v^\mu = \frac{d^2 x^\mu}{dt^2} + \Gamma^\mu_{\nu\lambda}\frac{dx^\nu}{dt}\frac{dx^\lambda}{dt} = 0\,,\tag{1.19}$$

which describes geodesic motion on the henceforth curved spacetime manifold $\mathcal{M}$. With respect to the above mentioned (non-rotating) Newtonian frames of reference, we have $\Gamma^i_{00} = \Phi_{,i}$ and all other $\Gamma^\mu_{\nu\lambda} = 0$. Effectively, the gravitational potential Φ has been 'absorbed' through $\Omega^\mu_{\nu\lambda}$ into the geometry, i.e., into the affine structure of

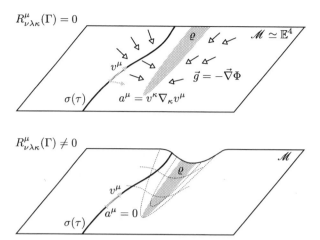

Fig. 1.7 Newtonian theory in the presence of gravity. *Upper*: Trajectory of a material (point) particle under the influence of an exact (gravitational) spatial vector force field $\mathbf{g} = -\boldsymbol{\nabla}\Phi$, according to the equations of motion in (1.13) and (1.16). Inertia is represented by a flat (integrable) affine structure or connection. *Lower*: Geodesic motion in a curved manifold, with gravity 'absorbed' into the geometry of spacetime, according to (1.19) with a curved (non-integrable) affine connection $\Gamma^{\mu}_{\nu\lambda} = \Lambda^{\mu}_{\nu\lambda} + \Omega^{\mu}_{\nu\lambda}$. The first graphic illustrates the standard or 'traditional' view of most textbook presentations, while the second depicts the alternative formulation of Newtonian theory in the geometric spirit of Einstein's general relativity. In both cases, the single source of gravity is the mass density ϱ and the gravitational potential is related to it by Poisson's equation

spacetime. The nonflat affine connection becomes a dynamical object depending on the gravitational potential and the mass density, and is now the basic element in this spacetime theory of gravity (Fig. 1.7). This has been masterfully phrased in Ehlers (1973b, p. 9f) and Ehlers (1973a, p. 78):

> Summarizing one may thus say that *the Eötvös–Dicke experiment suggests that the gravitational field is not a* (frame-independent) *vector field, but a non-integrable symmetric connection, whose geodesics are the free fall trajectories,* a conclusion which anticipates one of the main ingredients of Einstein's general theory.

> Whereas formally the local laws of Cartan's theory of spacetime, gravity, and dynamics [...], if expressed with respect to non-rotating coordinate frames, do not differ from those of standard Newtonian theory as given in the textbooks, conceptually it embodies an important advance by denying the separate existence of an integrable affine connection representing the inertial field and a vector field representing gravitation, and introducing instead of these two structures a single non-integrable connection representing both inertia and gravity. An empirically unjustifiable, fictious distinction has thereby been removed, and the true nature of gravity as a connection has been recognized.

The corresponding (nonvanishing) curvature tensor has components

$$R^{\mu}{}_{\nu\lambda\kappa}(\Gamma) = 2t_{\nu}s^{\mu\sigma}\,\Phi_{;\sigma;[\lambda}t_{\kappa]}\ ,$$

with the properties $t_{[\sigma} R^{\mu}{}_{\nu]\lambda\kappa} = 0$, and $s^{\mu\sigma} R^{\nu}{}_{\lambda\sigma\kappa} = s^{\nu\sigma} R^{\mu}{}_{\kappa\sigma\lambda}$, which in turn imply the existence of a potential Φ and a flat connection $\Lambda^{\mu}_{\nu\lambda}$ such that the nonflat connection satisfies $\Gamma^{\mu}_{\nu\lambda} = \Lambda^{\mu}_{\nu\lambda} + \Omega^{\mu}_{\nu\lambda}$, and $R^{\mu}{}_{\nu\lambda\kappa}(\Gamma)$ as stated above (see Trautman 1965, 1967). With respect to a Newtonian reference frame, we have $R^{i}{}_{0k0} = \Phi_{,ik}$ and $R^{\mu}{}_{\nu\lambda\kappa} = 0$ otherwise. Calculating the Ricci tensor, and noting (1.15), we now get a four-dimensional spacetime version of Poisson's equation, formulated first in Friedrichs (1928):

$$R_{\mu\nu}(\Gamma) = R^{\sigma}{}_{\mu\sigma\nu}(\Gamma) = t_{\mu\nu} s^{\sigma\kappa} \Phi_{;\sigma;\kappa} = 4\pi\varrho t_{\mu} t_{\nu} \ .$$

With this change in the affine structure of spacetime—called *Newton–Cartan theory* (or 'general-nonrelativistic spacetime' by Ehlers)—our alternative formulation of Newtonian theory under the influence of gravity is a collection $(\mathcal{M}, \nabla_{\mu}, dt, ds^2, \varrho)$ with a set of *local* field equations. We have a dynamical, linear and symmetric, affine, curved and non-integrable connection $\Gamma^{\mu}_{\nu\lambda}$, a pair of singular, symmetric metrics $(t_{\alpha\beta}, s^{\alpha\beta})$, and a scalar field ϱ specifying the distribution of matter [see also Misner et al. (1973, Box 12.4), and the 'geometrization lemma' in Malament (1986a, p. 191)], and these satisfy

$$t_{[\sigma} R^{\alpha}{}_{\kappa]\beta\gamma} = 0 \ , \tag{1.20}$$

$$s^{\alpha\sigma} R^{\kappa}{}_{\beta\sigma\gamma} = s^{\kappa\sigma} R^{\alpha}{}_{\gamma\sigma\beta} \ , \tag{1.21}$$

$$t_{\alpha;\gamma} = 0 \ , \tag{1.22}$$

$$s^{\alpha\beta}{}_{;\gamma} = 0 \ , \tag{1.23}$$

$$s^{\alpha\beta} t_{\alpha} = 0 \ , \tag{1.24}$$

$$R_{\alpha\beta} = 4\pi\varrho t_{\alpha} t_{\beta} \ . \tag{1.25}$$

Material particles in 'free fall', i.e., affected only by gravitational forces, follow geodesic motion according to (1.19) in a now curved manifold. Although the sections of absolute simultaneity—the spacelike hypersurfaces representing space—are still flat, or Euclidean, our four-dimensional spacetime geometry is essentially non-Euclidean. [According to Malament's spatial flatness proposition (Malament 1986a, p. 188), Poisson's equation in its geometrized formulation of (1.25) already implies *spatial flatness* in any classical spacetime model; see also Künzle (1972)]. Hence, inertial systems may no longer be characterized by the vanishing of the components of the affine connection. There simply does not exist a privileged class of exact global inertial coordinates with respect to which $\Gamma^{\mu}_{\nu\lambda} = 0$ in general. In the presence of gravity it is only possible to define local inertial frames. The local field equation (1.20) may be called with Ehlers the 'Newtonian restriction' or the 'law of existence of absolute rotation' (see Ehlers 1981, p. 72 and Ehlers 1991, Sect. 4). This condition ensures that our spacetime model really represents standard Newtonian theory proper, but not a more generalized non-relativistic gravitational theory. If we impose some additional, global conditions, e.g., the gravitational field and potential of an isolated source tend to zero far from the system (i.e., some suitable conditions of asymptotic *spatial* flatness, thereby replacing the 'Newtonian restriction' condition), the notion of a well defined inertial system (as in the absence of gravity) is restored asymptotically (see Trautman 1965, 1967). In particular, in Newtonian cosmology, where true 'isolated' or 'island' systems never exist (e.g., cosmological models with spatially homogeneous or asymptotically homogeneous matter distribution), these global assumptions may not be achievable.

We summarize our efforts to geometrize Newtonian gravitation theory and its significance by a quotation from Malament (1986a, p. 181) [see also the associated remarks in Malament (2012, p. 248f)]:

> First, it shows that various features of general relativity, once thought to be uniquely characteristic of it, do not distinguish it from Newtonian theory. On reformulation the latter, too, is a 'generally covariant' theory of four-dimensional space-time structure. As in general relativity, gravity is interpreted geometrically. Rather than thinking of particles as being deflected from straight trajectories by the presence of a gravitational potential, one thinks of them simply as traversing geodesic trajectories in curved space-time. What seemed a mysterious 'force' is now nothing but a manifestation of space-time curvature. Also as in general relativity, space-time becomes 'dynamical' under the reformulation. Rather than being a fixed, invariant backdrop against which physics unfolds, space-time affine structure participates in the unfolding. Curvature is dynamically correlated with the presence of matter according to Poisson's equation.
>
> Second, the work of Cartan, Friedrichs, and others clarifies the gauge status of the Newtonian gravitational potential. In the geometric formulation of Newtonian theory one works with a single (curved) affine structure. It can be decomposed into two pieces—a flat affine structure and a gravitational potential—to recover the standard formulation of the theory. But in the absence of special boundary conditions, the decomposition will not be unique. Physically, there is no unique way to divide into inertial and gravitational components the forces experienced by particles. Neither has any direct physical significance. Only their 'sum' does. (This is one version of the 'equivalence principle'.) It is an attractive feature of the geometric reformulation that it trades two gauge quantities for this sum.
>
> Third, the work under discussion provides the means with which to make clear geometric sense of the standard claim that Newtonian gravitational theory is the 'classical limit' of general relativity. One considers an appropriate one-parameter family of relativistic models $(M, g_{ab}(\lambda), T_{ab}(\lambda))$ satisfying Einstein's equation, defined for $\lambda > 0$, and then proves that in the limit as $\lambda \to 0$ a classical model $(M, t_a, h^{ab}, \nabla_a, \varrho)$ satisfying (the recast version of) Poisson's equation is defined. Intuitively, as $\lambda \to 0$, the null cones of $g_{ab}(\lambda)$ 'flatten' until they become degenerate.

The last item here refers to the fact that, within Ehlers' so-called frame theory (Ehlers 1981, 1991) for $\lambda = 0, G > 0$, Poisson's geometric field equation $R_{\alpha\beta} = 4\pi\varrho t_\alpha t_\beta$ is in a mathematically precise way the limit of Einstein's field equations. [In this 'classical limit' of general relativity, Newtonian gravitational theory implies, according to the spatial flatness proposition mentioned above, that space is flat (Malament 1986b)!] As Ehlers emphasized, such a rigorous formulation of a 'limit relation' between Newton's and Einstein's theories of spacetime and gravitation is especially important, and most satisfying, since although both theories are extremely successful in their range of validity—the latter superseding the former—they nevertheless rest upon radically different mathematical concepts and physical terminology. This even more so since another quite important transition, the transition from the (microscopic) quantum regime to the classical (macroscopic) 'world' is much less well understood, and as yet there is no real consensus on any kind of limit relation between classical and quantum physics.

Finally, the importance of the Newton–Cartan or general-nonrelativistic space-time model constructed above in physics is once again best exposed by Ehlers (1973a, p. 79) (see also Rosen 1972):

> All of nonrelativistic physics including quantum mechanics can be reformulated without difficulty within the framework of general-nonrelativistic spacetime. [...] All the non-gravitational local laws have, in local inertial frames, the same form as in the gravity-free special spacetime. Thus Einstein's strong principle of equivalence is incorporated satisfactorily (as far as slow motion, low energy phenomena are concerned) into nonrelativistic physics.

Although we were able to formulate Newtonian gravitation theory in a four-dimensional spacetime setting without recourse to a 'gravitational force', and while general relativity also ends up abandoning the concept of a force altogether, we now turn to a detailed synopsis of Newton's law of gravitation. Poisson's equation $\Delta\Phi_{\mathrm{grav}} = 4\pi\varrho$, the second order elliptic partial differential equation connecting the gravitational potential Φ_{grav} and the mass density ϱ (the source), is a local statement of Newton's action-at-a-distance law of gravitation with its characteristic inverse-square radial fall-off behaviour and quasi-infinite range:

$$(\mathbf{F}_{\mathrm{grav}})_{12}(\mathbf{r}) = -G\frac{(m_1)_{\mathrm{g}}(m_2)_{\mathrm{g}}}{r^2} \cdot \mathbf{e}_r \ . \tag{1.26}$$

Here, $r = |\mathbf{r}|$, G is the gravitational constant, $(m_1)_{\mathrm{g}}$ is the *passive* gravitational mass, $(m_2)_{\mathrm{g}}$ the *active* gravitational mass of the gravitating object (e.g., the Earth, the Sun, etc.), and $\mathbf{e}_r = \mathbf{r}/r$ is the unit vector in the direction of the line joining the center of the two massive bodies. Newton's great insight was to put forward the hypothesis that this force law is not limited to Earth-based systems (such as the fall of an apple towards the center of the Earth), but applies to all gravitationally acting massive bodies, and in particular the Solar System.

Newton's force law has some very distinctive characteristics, some of them already mentioned in Sect. 1.1, and which reflect certain facts of nature of amazing simplicity:

• Newton's gravitational force constitutes a *universal* force field involving all bodies in the universe. In principle, gravity cannot be shielded, in contrast to the way electrical forces can be shielded by a Faraday cage. (Any 'shielding device' would only bring in more gravity!) As a consequence of Einstein's $E = mc^2$ law, any physical system, which by definition has energy and therefore (gravitational) mass, participates both actively and passively in gravitation.

• It is a *central* force, i.e., it acts isotropically in all directions. Such isotropy of a physical vector quantity is manifestly the simplest possible attribute, and in a way the most 'natural' one between two bodies. Due to its central force character, for a mechanical system (like the Solar System or the Earth–Moon system) governed by Newton's second law and subject to (1.26), angular momentum is conserved. (As central forces appear quite often in nature, this conservation law for angular momentum has a kind of 'universal' validity.)

- The fall-off behaviour of the field strength, radial from the center of the active gravitating mass, has an *inverse-square* dependence, which results in a *long-range*, action-at-a-distance force (i.e., with an instantaneous effect). Compared to all other forces known in nature (except Coulomb's law, see below), Newton's gravitational force has the weakest possible fall-off behavior, which in turn implies that it is the force with maximum (mathematically infinite) range of influence. Furthermore, compared to other fundamental forces, the gravitational force is very weak, mainly because of the small value of the gravitational constant $G = 6.673\,84(80) \times 10^{-11}\,\mathrm{m^3\,kg^{-1}\,s^{-2}}$ (CODATA 2014). (Here, a rather 'extreme' example is provided by the electrical repulsion of two electrons, which is a factor 10^{42} stronger than their gravitational attraction!) Ranging from terrestrial attraction to the evolution of our universe, it is mainly on astrophysical and cosmological scales that the gravitational force becomes the predominant force governing nature (and all the more so because the universe is on average electrical neutral).
- The force is always *attractive*, due to the minus sign in (1.26), the fact that there do not exist masses with opposite sign, and the fact that all masses are positive (whereas there are positive and negative electrical charges, whence there can be electrical repulsion).
- Equation (1.26) is *linear* in the masses m_1 and m_2 and symmetric in the exchange of the two bodies. According to Newton's third law, the principle 'action equals reaction' $\mathbf{F}_{12} = -\mathbf{F}_{21}$ (which applies to all known forces), this symmetry has even more relevance. The linearity of Newton's law of gravitation in the masses may be completed by Newton's fourth law, which states the principle of linear superposition of mechanical forces. Geometrically, this is achieved by a parallelogram of forces. This principle is valid also for electrical and other forces, representing an enormous simplicity of nature with regard to a multitude of phenomena.
- Mathematically, the gravitational force is an *exact vector field*, as are in general all force fields within the class of central forces, i.e., of type

$$\mathbf{F}(\mathbf{r}, \dot{\mathbf{r}}) = f(r, \dot{r})\mathbf{r},$$

with the consequence that there exists a single function, namely the (negative) gravitational potential

$$\Phi_{\mathrm{grav}}(r) = -\frac{Gm_1 m_2}{r}, \quad \text{with } \mathbf{F}_{\mathrm{grav}} = -\nabla\Phi_{\mathrm{grav}}.$$

In essence, Newton's gravitational force is a mathematically 'simple' kind of force, conceivably even the simplest one. An immediate consequence of the $1/r$ potential is that, for Newton's law with force given by (1.26), the so-called

Runge–Lenz vector $\mathbf{A} = \mathbf{p} \times \mathbf{L} - m_1 m_2^2 G\, \mathbf{e}_r$ is another conserved quantity for the orbiting particle m_2, along with its angular momentum $\mathbf{L} = \mathbf{r} \times \mathbf{p}$.

- The physical concept of a *point mass* (described mathematically by a distribution) seems to be especially appropriate for the gravitational force, as the attraction of a spherically symmetric body (e.g., to a good approximation the Earth and, more importantly, virtually all stars) acts as though the whole mass of the body were concentrated at its center. This is a direct consequence of the $1/r$ behavior of its potential (and is true also for an exponentially decaying Yukawa potential $\Phi_{\text{Yukawa}}(r) \sim -\mathrm{e}^{-r/\lambda}/r$, as realized, e.g., by certain nuclear forces). Quite accidentally, due to the identical fall-off behavior of electric forces, the concept of a 'point charge' works equally well. Furthermore, if the body is a spherically symmetric massive (or charged) shell, no net gravitational (electric) force is exerted by the shell on any object in its interior (Newton's shell theorem).

It should be mentioned that the specific $1/r^2$ law of the gravitational force is the only one (besides an unphysical force law $\mathbf{F} \sim \mathbf{r}$ diverging at large distances) that would result in closed orbits for the planets in our Solar System, so among other things, it is one of the crucial features that actually made life possible on Earth!

Newton's law has striking similarities with a second fundamental force, namely the force law of electrostatics, Coulomb's law, describing the force between two charges e_1 and e_2 :

$$\mathbf{F}_{\text{el}}(\mathbf{r}) = \frac{1}{4\pi\varepsilon_0} \frac{e_1 e_2}{r^2} \mathbf{e}_r \ . \tag{1.27}$$

As for Newton's law, Coulomb's static inverse-square law is linear in the charges, but may also be repulsive, and it obeys Newton's 'action equals reaction' principle as well. Due to Gauss' law, Coulomb's law may also be rewritten as a potential equation, namely $\Delta\Phi_{\text{el}}(r) = -\varrho_{\text{c}}(r)/\varepsilon_0$, with the electric charge density ϱ_{c}. Nevertheless, it is well known that gravitation is physically and mathematically very different from electromagnetism (see, e.g., Ciufolini and Wheeler 1995, Sect. 7.1): although the two fundamental forces are almost identical in the static case (apart from a minus sign and some numerical coefficients) their dynamical behavior (in particular in the relativistic regime) differs as radically as one can imagine.

Over the last two decades, there has been a renewed interest in experimental tests of the inverse-square law (ISL), especially as regards Newton's law of gravitation [a detailed and updated presentation of the different methods used to test ISL, such as orbital, geophysical, and laboratory observations, and other modern experiments can be found in Ohanian and Ruffini (2013, Sect. 1.2)]. Historically, such tests of either Coulomb's or Newton's inverse-square law in the form

$$F_{\text{grav}}(r) = -G\frac{m_1 m_2}{r^{2+\varepsilon}} \tag{1.28}$$

were mainly carried out in order to explain observational violations of the given theory [e.g., in 1894, Hall found $\varepsilon = 1.6 \times 10^{-7}$ to fit the 'anomalous' perihelion advance of Mercury with Newton's theory of gravity (Hall 1894)]. However, as was noticed by Adelberger et al. (2003), a parametric ansatz according to (1.28) is rather inappropriate: "From the perspective of Gauss's Law, the exponent 2 [in (1.28)] is a purely geometrical effect of three space dimensions, so this parametrization was not well-motivated theoretically." More recently, laboratory measurements of potential ISL violations with significantly increased sensitivity [for an excellent review, see Adelberger et al. (2003, 2009) and references therein] try to set bounds on a possible Yukawa addition to the exact $1/r$ Newtonian potential, viz.,

$$\Phi_{\text{test}}(r) = -G\frac{m_1 m_2}{r}\left(1 + \alpha e^{-r/\lambda}\right) , \tag{1.29}$$

where α is a dimensionless strength parameter relative to Newtonian gravity and λ is a length scale or range related to the mass $\mu = \hbar/\lambda$ of the mediating particle in a relativistic field theoretic approach. (For gravity, this particle is the hypothetical spin-two graviton, and for electromagnetism the spin-one photon.) Testing any deviation of the ISL behavior of Newton's law (or Coulomb's law) is therefore equivalent to measuring a (tiny) nonzero value for the rest mass of a graviton (or photon) (see, e.g., Goldhaber and Nieto 2010).

Here, we only mention orbital observations based on laser- and radar-ranging data in the Solar System and Earth–Moon system on some characteristic length scales, and recent laboratory measurements on very small scales (compared to galactic or even greater cosmological dimensions of the universe, where deviations from the ISL character of Newton's law might be expected anyway). Precise planetary distance measurements by radar ranging to spacecraft orbiting Mercury, Venus, Mars, and Jupiter set limits on α for λ between 10^8 and 10^{12} m to $|\alpha| < 10^{-8}$. Lunar laser-ranging data restrict α by $|\alpha| < 10^{-10}$ for $\lambda \approx 10^8$ m. Satellite laser ranging of the two LAGEOS satellites (see Sect. 4.4) over 13 years set bounds $|\alpha| < 10^{-11}$ for $\lambda \approx 6$ km, i.e., at approximately one Earth radius (Lucchesi and Peron 2010). The (α, λ) parameter space for laboratory tests ranging from a few meters down to a few microns are plotted in an updated status review by Newman et al. (2009). Here, e.g., torsion balance experiments reveal $|\alpha| < 10^{-2}$ for $\lambda > 0.2$ mm [see also the recent measurements in this regime by Yang et al. (2012)].

After this synopsis of the characteristics of Newton's law of gravitation and the observational evidence for it, we come back to its local formulation in terms of Poisson's equation and consider the most basic visible objects to appear in our universe, the multitude of stars. There are several important and astrophysically relevant open questions. For instance, for which classes of equations of state $\varrho(p) \geq 0$ characterising such equilibrium configurations, and for which parameters such as the mass M and angular momentum J, do there exist solutions of the equations

$$\Delta \Phi = 4\pi\varrho , \tag{1.30}$$

$$\nabla p = -\varrho \nabla \Phi , \tag{1.31}$$

which describe stars of compact support? And what are the stability limits against rotational disruption or gravitational collapse? Poisson's equation complemented by Euler's equation actually amounts to solving a nonlinear, singular (at the star's surface where ϱ is zero), and free boundary value problem, i.e., a quite involved mathematical problem.

The mathematical formulation of static and rotating stars within this class of nonrelativistic equations is quite obvious, and the question of their solutions seems to be a rather 'natural' one, because the ubiquitous existence of stars is evident on astrophysical grounds (we shall return to the relativistic star problem in Sect. 3.3). However, a mathematical solution was sought only quite late, and it turned out to be more intricate than one would expect at first sight. For a long time, only those stars with (unrealistic) constant mass density and ellipsoidal figure and equipotential surfaces were treated, and by many famous mathematicians and physicists (Newton, Jacobi, MacLaurin, Dedekind). [For an overview of the existence of different types of self-gravitating equilibrium figures of rotating ideal fluid bodies see Chandrasekhar (1969) for Newtonian theory and Meinel et al. (2008, Sect. 3.3) for general relativity]. But for high angular velocities, even constant density stars can have other (even non-convex) figures. Therefore, beyond the first 'bifurcation point', uniqueness of solutions cannot be expected.

For Newtonian theory it was relatively easily proven in Carleman (1918) and in Lichtenstein (1928) that static ideal fluid bodies are necessarily spherical. Concerning a general existence proof for rotating ideal fluid stars, only the method of Auchmuty and Beals (1971) is available. This is based on variational techniques on very special function spaces, and applies only to a quite reduced class of stars. A generalization of a standard method of proof (using the Banach fixed point theorem) from static stars (Schaudt 2000) to rotating stars has not yet succeeded. [In the degenerate case, rotating dust disks do exist in Newtonian gravity (see, e.g., Binney and Tremaine 1994, Sect. 5.3), and also in Einsteinian gravity (Neugebauer and Meinel 1994).] However, it has long been known, and using quite different methods of proof, that isolated Newtonian rotating dust stars (with pressure p identically zero) cannot exist (see Bonnor 1977). [Nevertheless, in Schaudt and Pfister (2001), it was shown by an explicit example that a rotating dust star can be stabilized by exterior (strained) matter which can be arbitrarily far from the dust region. This makes it especially clear that the star problem is a genuinely global one.]

In general, the problem of the existence of solutions for rotating stars remains essentially unsolved in nonrelativistic Newtonian gravity, as it does also in relativistic Einsteinian gravity, and this even for the 'simplest' cases of rotating stars in complete equilibrium, which rotate in a stationary, axisymmetric, and rigid manner, and behave like an ideal fluid (see Sect. 3.3). However, for a static and spherically symmetric fluid star, an existence proof for quite general equations of state was established in Pfister (2011). If rotating dust stars existed in general relativity, they would represent examples of a rather improbable perfect balance between the attractive quasi-Newtonian force (gravitoelectricity) and the repulsive gravitomagnetism (see Pfister 2010 and Sect. 4.4). For the state of the art regarding results in Newtonian gravity, see, e.g., Li (1991), and for a recent and fully comprehensive overview on rotating stars in Einsteinian gravity, see Friedman and Stergioulas (2013).

From this negative perspective, or positive depending on one's point of view, more than 325 years after the foundations of Newton's ingenious concepts and laws in mechanics and gravitation which resulted in his *Principia*, Newtonian physics is still a fascinating and open field of activity of physical science.

References

Adelberger, E.G., Heckel, B.R., Nelson, A.E.: Tests of the gravitational inverse-square law. Annu. Rev. Nucl. Part. Sci. **53**, 77–121 (2003)

Adelberger, E.G., et al.: Torsion balance experiments: a low-energy frontier of particle physics. Prog. Part. Nucl. Phys. **62**, 102–134 (2009)

Auchmuty, J.F.G., Beals, R.: Variational solutions of some nonlinear free boundary problems. Arch. Ration. Mech. Anal. **43**, 255–271 (1971)

Barbour, J.B.: Absolute or Relative Motion, Vol. 1: The Discovery of Dynamics. Cambridge University Press, Cambridge (1989)

Barbour, J.B.: Einstein and Mach's principle. In: Renn, J. (ed.) The Genesis of General Relativity, vol. 3, pp. 569–604. Springer, Dordrecht (2004)

Binney, J., Tremaine, S.: Galactic Dynamics. Princeton University Press, New Jersey (1994)

Bonnor, W.B.: A rotating dust cloud in general relativity. J. Phys. A **10**, 1673–1678 (1977)

Brown, H.R.: Physical Relativity. Oxford University Press, Oxford (2005)

Carleman, T.: Über eine isoperimetrische Aufgabe und ihre physikalischen Anwendungen. Math. Zeitschr. **3**, 1–7 (1918)

Cartan, E.: Sur les varietes a connexion affine et la théorie de la relativité généralisée. Ann. Ec. Norm. Sup. **40**, 325–412 (1923); **41**, 1–15 (1924)

Castagnino, M.: Some remarks on the Marzke–Wheeler method of measurement. Nuovo Cimento B **54**, 149–150 (1968)

Chandrasekhar, S.: Ellipsoidal Figures of Equilibrium. Yale University Press, Connecticut (1969)

Ciufolini, I., Wheeler, J.A.: Gravitation and Inertia. Princeton University Press, Princeton (1995)

CODATA Recommended Values. National Institute of Standards and Technology (2014)

Cohen, I.B.: Newton's discovery of gravity. Sci. Am. **244/3**, 122–133 (1981)

Earman, J.: World Enough and Space-Time: Absolute versus Relational Theories of Space and Time. MIT Press, Cambridge (1989). Esp. Chap. 2 Classical Space-Times

Earman, J., Friedman, M.: The meaning and status of Newton's law of inertia and the nature of gravitational forces. Philos. Sci. **40**, 329–359 (1973)

Ehlers, J.: The nature and structure of spacetime. In: Mehra, J. (ed.) The Physicist's Conception of Nature, pp. 71–91. Reidel, Dordrecht (1973a)

Ehlers, J.: Survey of general relativity theory. In: Israel, W. (ed.) Relativity, Astrophysics and Cosmology, pp. 1–125. Reidel, Dordrecht (1973b). Esp. Sect. 1.2 Newtonian space-time, Mechanics, and Gravity Theory

Ehlers, J.: Über den Newtonschen Grenzwert der Einsteinschen Gravitationstheorie. In: Nitsch, J., et al. (eds.) Grundlagenprobleme der modernen Physik, pp. 65–84. Bibliographisches Institut, Mannheim (1981)

Ehlers, J.: The Newtonian limit of general relativity. In: Ferrarese, G. (ed.) Classical Mechanics and Relativity: Relationship and Consistency, pp. 95–106. Napoli-Bibliopolis, Napoli (1991)

Ehlers, J.: Space-time structures. In: Krüger, L., Falkenburg, B. (eds.) Physik, Philosophie und die Einheit der Wissenschaften, pp. 165–176. Akademie-Verlag GmbH, Heidelberg (1995)

Einstein, A.: Prinzipielles zur allgemeinen Relativitätstheorie. Ann. Physik **55**, 241–244 (1918)

Einstein, A.: Inwiefern läßt sich die moderne Gravitationstheorie ohne die Relativität begründen? Naturwissenschaften **8**, 1010–1011 (1920)

Einstein, A.: The Meaning of Relativity. Methuen, London (1922)

Einstein, A.: Newton's Mechanik und ihr Einfluß auf die Gestaltung der theoretischen Physik. Die Naturwissenschaften **15**, 273–276 (1927)

Feynman, R.P., Leighton, R.B., Sands, M.: The Feynman Lectures of Physics, vol. I. Addison-Wesley, Reading (1965)

Frege, G.: Über das Trägheitsgesetz. Zeitschr. f. Philosophie u. philosoph. Kritik **98**, 145–161 (1891)

Friedman, M.: Foundations of Space-Time Theories. Princeton University Press, Princeton (1983). Reprinted in the Princeton Legacy Library as paperback edition (2014)

Friedman, J.L., Stergioulas, N.: Rotating Relativistic Stars. Cambridge University Press, Cambridge (2013)

Friedrichs, K.: Eine invariante Formulierung des Newtonschen Gravitationsgesetzes und des Grenzübergangs vom Einsteinschen zum Newtonschen Gesetz. Math. Ann. **98**, 566–575 (1928)

Galilei, G.: Dialogue Concerning the Two Chief World Systems—Ptolemaic & Copernican (1630). Translated by S. Drake. University of California Press, Berkeley (1967)

Giulini, D.: Das Problem der Trägheit. Philos. Nat. **39**, 343–374 (2002)

Giulini, D.: The rich structure of Minkowski space. In: Petkov, V. (ed.) Minkowski Spacetime: A Hundred Years Later, pp. 83–132. Springer, Berlin (2009)

Giulini, D.: Instants in physics—point mechanics and general relativity. In: von Müller, A., Filk, T. (eds.) Rethinking Time at the Interface of Physics and Philosophy, 278 pp. Springer, Berlin (2015)

Goldhaber, A.S., Nieto, M.M.: Photon and graviton mass limits. Rev. Mod. Phys. **82**, 939–979 (2010)

Gundlach, J.H., et al.: Laboratory test of Newton's second law for small accelerations. Phys. Rev. Lett. **98**, 150801 (2007)

Hall, A.: A suggestion in the theory of Mercury. Astron. J. **14**, 49–51 (1894)

Havas, P.: Four-dimensional formulations of Newtonian mechanics and their relations to the special and general theory of relativity. Rev. Mod. Phys. **36**, 938–965 (1964)

Havas, P.: Foundation problems in general relativity. In: Bunge, M. (ed.) Delaware Seminar in the Foundations of Physics, vol. 1, pp. 124–148. Springer, Berlin (1967)

Hilbert, D.: Grundlagen der Geometrie. Teubner, Leipzig (1899)

Kadison, L., Kromann, M.T.: Projective Geometry and Modern Algebra. Birkhäuser, Boston (1996)

Klein, F.: Vergleichende Betrachtungen über neuere geometrische Forschungen. A. Deichert, Erlangen (1872); Reprinted in Mathematische Annalen (Leipzig) **43**, 43–100 (1892)

König, A.: Über die neueren Versuche zu einer einwurfsfreien Grundlegung der Mechanik. Verhandlungen d. Physikal. Ges. Berlin **5**, 73–74 (1886)

Kretschmann, E.: Über die prinzipielle Bestimmbarkeit der berechtigten Bezugssysteme beliebiger Relativitätstheorien. Ann. Phys. (Leipzig) **48**, 907–942 (1915)

Kretschmann, E.: Über den physikalischen Sinn der Relativitätspostulate, A. Einsteins neue und seine ursprüngliche Relativitätstheorie. Ann. Physik **53**, 575–614 (1917)

Kriele, M.: Spacetime—Foundations of General Relativity and Differential Geometry. Springer, Berlin (1999)

Künzle, H.P.: Galilei and Lorentz structures on space-time: comparison of the corresponding geometry and physics. Ann. Inst. Henri Poincaré Sec. A **42**, 337–362 (1972)

Lange, L.: Über die wissenschaftliche Fassung des Galilei'schen Beharrungsgesetzes. In: v. W. Wundt (Hrsg.) Philosophische Studien, Bd. II, pp. 266–297 (1885a)

Lange, L.: Über das Beharrungsgesetz. Berichte über Verhandlungen der Königlich Sächsischen Gesellschaft der Wissenschaften, pp. 333–351. Mathematisch-physikalische Klasse, Leipzig (1885b). English translation in Eur. Phys. J. H **39**, 251–262 (2014)

Lange, L.: Die geschichtliche Entwicklung des Bewegungsbegriffs und ihr voraussichtliches Endergebnis. W. Engelmann, Leipzig (1886)

Lange, L.: Das Inertialsystem vor dem Forum der Naturforschung. In: v. W. Wundt (Hrsg.) Philosophische Studien, Bd. XX, pp. 1–71 (1902)

Laugwitz, D.: Differential and Riemannian Geometry. Academic, New York (1965)

Li, Y.Y.: On uniformly rotating stars. Arch. Ration. Mech. Anal. **115**, 367–393 (1991)

Lichtenstein, L.: Über eine Eigenschaft der Gleichgewichtsfiguren rotierender Flüssigkeiten, deren Teilchen einander nach dem Newtonschen Gesetze anziehen. Math. Zeitschr. **28**, 635–640 (1928)

Lucchesi, D.M., Peron, R.: Accurate measurement in the field of the Earth of the general-relativistic precession of the LAGEOS II pericenter and new constraints on non-Newtonian gravity. Phys. Rev. Lett. **105**, 231103 (2010)

Mach, E.: Die Geschichte und die Wurzel des Satzes von der Erhaltung der Arbeit. Calve, Prag (1872)

Mach, E.: Die Mechanik in ihrer Entwicklung. Historisch-kritisch dargestellt. Brockhaus, Leipzig (1883)

Malament, D.: Newtonian gravity, limits, and the geometry of space. In: Colodny, R.G. (ed.) From Quarks to Quasars: Philosophical Problems of Modern Physics, pp. 181–201. University of Pittsburgh Press, Pittsburgh (1986a)

Malament, D.: Gravity and spatial geometry. In: Marcus, R.B., et al. (eds.) Studies in Logic and the Foundations of Mathematics, vol. 114, pp. 405–411. Elsevier, Amsterdam (1986b)

Malament, D.: Topics in the Foundations of General Relativity and Newtonian Gravitation Theory. University of Chicago Press, Chicago (2012)

Marion, J.B.: Classical Dynamics of Particles and Systems. Academic, New York (1965). Fourth edition by J.B. Marion, and S.T. Thornton, Harcourt Brace, Fort Worth (1995)

Marsden, J.E., Ratiu, T.S.: Introduction to Mechanics and Symmetry, 2nd edn. Springer, New York (1999)

Marzke, R.F., Wheeler, J.A.: Gravitation as geometry. I. The geometry of space-time and the geometrodynamical standard meter. In: Chiu, H.-Y., Hoffman, W.F. (eds.) Gravitation and Relativity, pp. 40–64. Benjamin, New York (1964)

Meinel, R., Ansorg, M., Kleinwächter, A., Neugebauer, G., Petroff, D.: Relativistic Figures of Equilibrium. Cambridge University Press, Cambridge (2008)

Misner, C.W., Thorne, K.S., Wheeler, J.A.: Gravitation. Freeman, San Francisco (1973)

Neugebauer, G., Meinel, R.: General relativistic gravitational field of a rigidly rotating disk of dust: axis potential, disk metric, and surface mass density. Phys. Rev. Lett. **73**, 2166–2168 (1994)

Neumann, C.: Über die Principien der Galilei-Newton'schen Theorie. Teubner, Leipzig (1870)

Newman, R.D., Berg, E.C., Boynton, P.E.: Tests of the gravitational inverse square law at short ranges. Space Sci. Rev. **148**, 175–190 (2009)

Newton, I.: Mathematical Principles of Natural Philosophy (1687). Translated and edited by I.B. Cohen and A. Whitman. University of California Press, Berkeley (1999)

Newton, I.: Four Letters from Sir Isaac Newton to Doctor Bentley Containing Some Arguments in Proof of a Deity. R. & J. Dodsley, London (1756)

Ohanian, H.C., Ruffini, R.: Gravitation and Spacetime, 3rd edn. Cambridge University Press, Cambridge (2013)

Penrose, R.: An Analysis of the Structure of Space-Time, Essay, Awarded the Adams Prize 1965–1966, pp. 1–132. Princeton University Press, Princeton (1967). Reprinted in Penrose, R., Collected Works, vol. 1, pp. 1953–1967. Oxford University Press, Oxford (2011), Chap. 28

Penrose, P.: Structure of space-time. In: DeWitt, C.M., Wheeler, J.A. (eds.) Battelle Rencontres: 1967 Lectures in Mathematics and Physics, pp. 121–235. Benjamin, New York (1968). Reprinted in Penrose, R., Collected Works, vol. 2, pp. 1968–1975. Oxford University Press, Oxford (2011), Chap. 36

Petzoldt, J.: Die Gebiete der absoluten und der relativen Bewegung. Annalen d. Naturphilosophie **7**, 29–62 (1908)

Pfister, H.: Newton's first law revisited. Found. Phys. Lett. **17**, 49–64 (2004)

Pfister, H.: Do rotating dust stars exist in general relativity? Class. Quantum Gravity **27**, 105016 (2010)

Pfister, H.: A new and quite general existence proof for static and spherically symmetric perfect fluid stars in general relativity. Class. Quantum Gravity **28**, 075006 (2011)

Pfister, H.: Ludwig Lange on the law of inertia. Eur. Phys. J. H **39**, 245–250 (2014)

Rosen, G.: Galilean invariance and the general covariance of nonrelativistic laws. Am. J. Phys. **40**, 683–687 (1972)

Schaudt, U.M.: On static stars in Newtonian gravity and Lane–Emden type equations. Ann. Inst. Henri Poincaré **1**, 945–976 (2000)

Schaudt, U.M., Pfister, H.: Isolated Newtonian dust stars are unstable but can be stabilized by exterior matter. Gen. Relativ. Gravit. **33**, 719–737 (2001)

Schlamminger, S., et al.: Test of the equivalence principle using a rotating torsion balance. Phys. Rev. Lett. **100**, 041101 (2008)

Schrödinger, E.: Space-Time Structure. Cambridge University Press, Cambridge (1950)

Schulmann, R., et al.: The Collected Papers of Albert Einstein, vol. 8. Princeton University Press, Princeton (1998)

Schwarzschild, K.: Was in der Welt ruht. Die Zeit (Vienna) **11/142**, 514–521 (1897). English translation in The Genesis of general relativity (ed. by J. Renn) vol. 3, pp. 183–189. Springer, Dordrecht (2004)

Seeliger, H.: Über das Beharrungsgesetz. Die geschichtliche Entwicklung des Bewegungsbegriffs. Vierteljahrschrift d. Astron. Ges. **22**, 252–259 (1887)

Seeliger, H.: Über die sogenannte absolute Bewegung. Sitzb. d. Königl. Bayer. Akad. d. Wiss, math-phys. Kl. **36**, 85–137 (1906)

Stevenson, F.W.: Projective Planes. Freeman, San Francisco (1972)

Straumann, N.: Klassische Mechanik. Lecture Notes in Physics, vol. 289. Springer, Berlin (1987). Second enlarged edition published as Theoretische Mechanik Springer, Berlin (2015)

Streintz, H.: Die physikalischen Grundlagen der Mechanik. Teubner, Leipzig (1883)

Thomson, J.: On the law of inertia; the principle of chronometry; and the principle of absolute clinural rest, and of absolute rotation. Proc. R. Soc. Edinb. 568–578 (1883a) [Session 1883–1884]

Thomson, J.: A problem on point-motions for which a reference-frame can so exist as to have the motions of the points, relative to it, rectilinear and mutually proportional. Proc. R. Soc. Edinb. 730–745 (1883b) [Session 1883–1884]

Thomson, W. (Lord Kelvin), Tait, P.G.: Treatise on Natural Philosophy. Cambridge University Press, Cambridge (1879)

Torretti, R.: Relativity and Geometry. Pergamon Press, Oxford (1983). Reprinted by Dover publication as paperback edition (1996)

Trautman, A.: Foundations and current problems of general relativity. In: Deser, S., Ford, K.W. (eds.) Lectures on General Relativity, 1964 Brandeis Summer Institute in Theoretical Physics, vol. 1, pp. 1–248. Englewood Cliffs, Prentice-Hall (1965). Esp. Chap. 5. Theories of Space, Time and Gravitation

Trautman, A.: Comparison of Newtonian and relativistic theories of space-time. In: Hoffman, B. (ed.) Perspectives in Geometry and Relativity, pp. 413–425. Indiana University Press, Bloomington (1967)

von Laue, M.: Die Relativitätstheorie, Bd. 1. Vieweg, Braunschweig (1911)

von Laue, M.: Dr. Ludwig Lange 1863–1936 (Ein zu Unrecht Vergessener). Die Naturwissenschaften **35**, 193–196 (1948)

Voss, A.: Die Prinzipien der rationellen Mechanik. Enzykl. d. Math. Wissensch. IV, 1. Teubner, Leipzig (1901)

Weyl, H.: Raum, Zeit, Materie, 1st edn. Springer, Berlin (1919); 5th edn. Springer, Berlin (1922); 8th edn. Springer, Berlin (1993)

Weyl, H.: Mathematische Analyse des Raumproblems. Dritte Vorlesung, pp. 14–22. Springer, Berlin (1923)

Weyl, H.: Geometrie und Physik. Die Naturwissenschaften **19**, 49–58 (1931)

Wigner, E.: The unreasonable effectiveness of mathematics in the natural sciences. Comm. Pure Appl. Math. **13**, 1–14 (1960)

Wohlwill, E.: Die Entdeckung des Beharrungsgesetzes, Zs. f. Völkerpsychologie u. Sprachwissenschaften **14**, 365–410 (1883)

Wohlwill, E.: Die Entdeckung des Beharrungsgesetzes. Zs. f. Völkerpsychologie u. Sprachwissenschaften **15**, 70–135, 337–387 (1884)

Yang, S.-Q., et al.: Test of the gravitational inverse square law at millimeter ranges. Phys. Rev. Lett. **108**, 081101 (2012)

Chapter 2
Free Particles and Light Rays as Basic Elements of General Relativity

2.1 The Seeds Disseminated by Hermann Weyl

In the first half of the twentieth century, the deepest and most innovative analysis of space and spacetime (after B. Riemann's ingenious foundation of the concept of a Riemannian manifold in his famous habilitation treatise *Über die Hypothesen, welche der Geometrie zu Grunde liegen*, see, e.g., Jost 2013) was certainly the one achieved by H. Weyl. He felt challenged by Einstein's general relativity, which showed that Riemannian geometry is not just a nice abstract mathematical model for a metric manifold, but is explicitly realized in physical spacetime. (Weyl also wrote the first textbook on general relativity, as early as the spring of 1918. For Weyl's comprehensive contributions to Einstein's theory, see, e.g., Ehlers 1985).

One question that was of particular interest to Weyl in this context was the so-called space problem (Raumproblem), which asks how Riemann's quadratic local metric $ds^2 = g_{ik}(x^a)dx^i dx^k$ is singled out from fourth-order or higher-order metrics like $ds^4 = h_{ikjl}(x^a)dx^i dx^k dx^j dx^l$ (already mentioned by Riemann), or even more general (Finslerian) metrics $ds^2 = g_{ik}(x^a, dx^b)dx^i dx^k$. [In Weyl's own words (Weyl 1922), this was the question of "the uniqueness of the Pythagorean determination of a measure".] A first answer to this question was already given in Helmholtz (1868): if infinitesimal rigid bodies can be freely rotated around arbitrary points, the space is Riemannian. But Weyl was not satisfied with this type of argument and sought in Weyl (1922) "to understand the metric nature of space from as simple and fundamental reasons as possible", in particular, using the linear affine connection (Levi-Civita's infinitesimal parallel transport) which comes with any homogeneous metric space, and which then applies to any signature of the metric. Weyl finally succeeded in proving (Weyl 1922) that only for a Riemannian metric does such a metric-compatible affine connection exist, and that it is then unique. However, he denounced his own proof as "mathematical rope-dancing". Today, of course, there exist simpler proofs of this fundamental fact about metric spaces (see, e.g., Laugwitz 1965, Sect. 15).

© Springer International Publishing Switzerland 2015
H. Pfister, M. King, *Inertia and Gravitation*, Lecture Notes in Physics 897,
DOI 10.1007/978-3-319-15036-9_2

For our approach to general relativity in this chapter, even more important is Weyl's extraction of a projective geometry and a conformal geometry from a (pseudo-)Riemannian manifold. This is carried out in detail in Weyl (1921). Projective geometry as such was of course already known and studied long before by Steiner, Plücker, Hilbert, and others. What is new is the connection with the paths of free particles, according to Einstein's equivalence principle. In Weyl's words:

> The inertial tendency of the world direction of a moving material particle, which enforces upon it a certain 'natural motion' if released into some world direction, is this unity of inertia and gravitation, which Einstein put in the place of both.

Weyl then considers projective transformations (which leave the inertial structure of the free particle paths invariant, but typically change other metric properties), and finds that these change the affine connections (Christoffel symbols) according to

$$\delta \Gamma^{\mu}_{\nu\lambda} = \delta^{\mu}_{\nu}\psi_{\lambda} + \delta^{\mu}_{\lambda}\psi_{\nu} \;,$$

with arbitrary functions $\psi_{\lambda}(x)$. He also defines a projective curvature tensor and finds that this is nontrivial only in manifolds of dimension $d \geq 3$. The conditions for a manifold to be projectively flat (i.e., all particle paths can globally be chosen as straight lines) are given explicitly, but they are different for $d = 2$ and $d \geq 3$.

Weyl's conformal geometry is an essentially new structure which was not considered before because attention was usually confined to positive-definite metrics (Weyl 1921):

> Characteristic for the conformal structure of a metric space is the infinitesimal cone of null directions $g_{ik}dx^{i}dx^{k} = 0$ belonging to each place. [...] The infinitesimal cone accomplishes in the neighborhood of a world point the separation of past and future; the conformal structure is the action connexion of the world which determines which world points are in a possible causal connection.

Weyl then considers conformal transformations which leave the null cone invariant but change the affine and therefore the inertial structure according to

$$\delta \Gamma^{\mu}_{\nu\lambda} = \frac{1}{2}(\delta^{\mu}_{\nu}\varphi_{\lambda} + \delta^{\mu}_{\lambda}\varphi_{\nu} - g_{\nu\lambda}\varphi^{\mu}) \;,$$

with arbitrary functions $\varphi_{\lambda}(x)$. He derives the following significant fact for physics (Weyl 1921): "The projective and conformal structure of a metric space determine its metric uniquely." This has the consequence that "solely by observing the 'natural' motion of material particles and action propagation, in particular, light propagation, the world metric can be determined; rulers and clocks are not necessary for that". It should be stressed, however, that this is only true if a Riemannian metric structure is already presupposed. Otherwise the projective and conformal structure do not have to be compatible, and even if they are, this leads to a Riemannian metric only under additional conditions, as we shall see in detail in Sect. 2.4.

The conformal structure leads again to a characteristic curvature tensor which Weyl had already derived earlier in Weyl (1918a), and which in the notation of Hawking and Ellis (1973, p. 41) reads:

$$C_{\mu\nu\lambda\sigma} = R_{\mu\nu\lambda\sigma} - \frac{1}{d-2}(g_{\mu\lambda} R_{\nu\sigma} + g_{\nu\sigma} R_{\mu\lambda} - g_{\mu\sigma} R_{\nu\lambda} - g_{\nu\lambda} R_{\mu\sigma})$$

$$+ \frac{1}{(d-1)(d-2)}(g_{\mu\lambda} g_{\nu\sigma} - g_{\mu\sigma} g_{\nu\lambda}) R \,,$$

where (in dimension d) $R_{\mu\nu\lambda\sigma}$ is the Riemann curvature tensor, $R_{\nu\sigma} = R^{\mu}_{\nu\mu\sigma}$ is the Ricci tensor, and $R = R^{\mu}_{\mu}$ is the curvature scalar. The tensor $C_{\mu\nu\lambda\sigma}$ is today called the Weyl tensor. It completely characterizes the vacuum part of a physical spacetime, e.g., the exterior of an isolated star or black hole, and gravitational waves. Conformal transformations have great importance for the global analysis of physical spacetimes, e.g., in the hands of R. Penrose, for compactifying spacetimes and defining the asymptotic lightlike manifold Scri (Penrose 1964), where gravitational waves can be read off (Frauendiener and Friedrich 2002), and for the important singularity theorems of general relativity (Penrose 1965). There are even proposals (namely Penrose's Weyl curvature hypothesis, see Sect. 3.3) to characterize the beginning of the universe (the big bang) and the irreversible evolution of the 'thermodynamical world system' by special properties of the Weyl tensor (Penrose 1979). Another interesting discovery derived in Weyl (1918a) is the fact that the Weyl tensor is nonzero only in dimensions $d \geq 4$, and that there are differences in the characterization of conformally flat manifolds (where the light cones are globally straight) for $d = 3$ and $d \geq 4$. (Manifolds with spacetime dimension below four would not admit sufficiently rich and interesting physical phenomena!)

A further important seed shown by Weyl in 1918 (Weyl 1918a), and in his famous textbook on general relativity entitled *Space, Time, Matter* (Weyl 1919, Sects. 17 and 40), which exists in several very different editions, concerns the idea of a gauge principle. In the original conception, this was based on the hypothesis that vector transport around a closed loop not only changes the direction of the vector (as is typically the case in a Riemannian manifold), but also its length. This was then applied to a geometrization of electrodynamics, and hence to a kind of unification of electrodynamics with gravity. However, the idea was immediately criticized, in particular by Einstein, and subsequently lay dormant for a long time. But as is well known, the gauge principle was restored to life and is now successfully applied in modern quantum field theory and elementary particle physics. On the other hand, concerning the present enthusiasm for a possible understanding of the generation of the masses of the elementary particles by the Higgs mechanism, it may be appropriate to quote a warning in Weyl (1931): "No good theory will account for the nature of mass without referring to gravitation."

We would like to close this section with another daring quotation from Weyl in a footnote to Weyl (1918a): "I am bold enough to believe that the totality of the physical phenomena can be derived from a single universal world law of the highest mathematical simplicity." Even 95 years later, this vision has obviously not yet been fulfilled.

2.2 Light Rays Define the Conformal Structure (Ehlers, Pirani, Schild)

For their seminal paper entitled *The geometry of free fall and light propagation* (Ehlers et al. 1972) the authors (EPS) were undoubtedly stimulated by Weyl's work, as presented in the previous section. [Within the more comprehensive context of classical, non-relativistic and relativistic spacetime models and structures, Ehlers has given a detailed synopsis of EPS axiomatics in Ehlers (1973b, Sect. 2), and with some further remarks in Ehlers (1973a, Sect. 4). For a non-technical introduction to this theme, in a rather descriptive style but physically precise language, see Ehlers (1988).] However, in a certain sense, they invert the logic in the analysis of the fundamental structure of physical spacetime (Ehlers et al. 1972, p. 64f):

> If, however, one wishes to give a *constructive* set of axioms for relativistic space-time geometries, which is to exhibit as clearly as possible the physical reasons for adopting a particular structure and which indicates alternatives, then the chronometric approach [e.g., by J. L. Synge in his book *Relativity: The Special Theory* (1956)] does not seem to be particularly suitable for the following three reasons. It seems difficult to derive *from the behaviour of clocks alone*, without the use of light signals, the *Riemannian* form of the separation, $ds^2 = g_{ij}dx^i dx^j$, rather than some other, first-degree homogeneous, functional form in the dx^i (as, for instance, the Newtonian form $ds = g_i dx^i$). Postulating this form axiomatically, one foregoes the possibility of understanding the reason for its validity. The second difficulty is that if the g_{ij} are defined by means of the chronometric hypothesis, it seems not at all compelling—if we disregard our knowledge of the full theory and try to construct it from scratch—that these chronometric coefficients should determine the behaviour of freely falling particles and light rays, too. Thus the geodesic hypotheses, which are introduced as additional axioms in the chronometric approach, are hardly intelligible; they fall from heaven [like the equation for ds^2 above]. Finally, once the geodesic hypotheses have been accepted, it is possible, in the theories of both special and general relativity, to construct clocks by means of freely falling particles and light rays, as shown by Marzke and, differently, by Kundt and Hoffmann. Thus, these hypotheses alone already imply a physical interpretation of the metric in terms of time. The chronometric axiom then appears either as redundant or, if the term 'clock' is interpreted as 'atomic clock', as a link between macroscopic gravitation theory and atomic physics: it claims the equality of gravitational and atomic time. It may be better to test this equality experimentally or to derive it eventually from a theory that embraces both gravitational and atomic phenomena, rather than to postulate it as an axiom.

Today, the equality of gravitational and atomic time is confirmed with a precision reaching 10^{-15}, because for millisecond pulsars the stability of the pulse period is similar to that achieved by the best terrestrial atomic clocks (Hobbs et al. 2012). (On short timescales atomic clocks even reach a precision of 10^{-18}.)

Together with the proportionality of these two clock types with 'clocks' functioning
on the basis of elastic forces, electromagnetic forces, strong interactions, weak
interactions (radioactive age determination), and possibly other processes, this
represents presumably the most convincing experimental sign that the goal of all
theoretical physicists to unify all physical phenomena of nature in one theory is no
idle dream! In Ehlers et al. (1972), we read:

> For these reasons, we reject clocks as basic tools for setting up the space-time geometry and
> propose to use light rays and freely falling particles instead. We wish to show how the full
> space-time geometry can be synthesized from a few assumptions about light propagation
> and free fall.

If the foundations of general relativity are gradually built up in this way from
light rays and free particles, then experimental tests of this theory can immediately
be ordered according to the question as to which parts of general relativity they
really test (e.g., the equivalence principle), and which other parts they leave open
(e.g., the Einstein field equations). Furthermore, such a gradual approach to general
relativity provides a natural way to perform modifications of general relativity by
reducing or extending its metric structure, leading to non-metric theories, gauge
theories, super-theories, etc.

Before one can analyze in detail and in a mathematical language the structures of
light rays and free particles in a physical spacetime $\mathcal{M}$, one has to fix the topological
structure of $\mathcal{M}$, as do EPS in Ehlers et al. (1972, Sect. 2, pp. 70–72). (See also
typical textbooks on differential topology, like Abraham et al. 1988.) $\mathcal{M}$ consists of
all events, i.e., the most elementary and most local acts of physical observation, like
the collision of two 'pointlike' particles. Although the totality of really observable
events is of course finite, it has proved reasonable and without contradiction (so
far!) to idealize $\mathcal{M}$ as a continuous and differentiable manifold, with the advantage
that the powerful methods of calculus are then applicable. The topological model
of this differentiable manifold which has proved good throughout physics can be
characterized as locally similar to four-dimensional Euclidean space $\mathbb{R}^4$. [Although
this wrongly suggests a locally isotropic four-dimensional space-time, attempts
to incorporate the light cone structure into the topology in the case of the so-
called Zeeman topologies (Zeeman 1967) have not really prevailed in mathematical
physics.]

The manifold $\mathcal{M}$ is therefore assumed to be completely covered by (finitely or
infinitely many) submanifolds $\mathcal{U}_\alpha$, each of which can be mapped by a uniquely
invertible chart Φ_α to an open submanifold of $\mathbb{R}^4$, where in the common region
$\mathcal{U}_\alpha \cap \mathcal{U}_\beta$, the charts Φ_α and Φ_β are connected by a diffeomorphism (C^3 is usually
sufficient). The manifold $\{\mathcal{U}_\alpha, \Phi_\alpha\}$ is then called an atlas or a coordinatization of
$\mathcal{M}$. A concrete realization of such a coordinatization can be given by so-called *radar
coordinates*: two observers who move on arbitrary (not necessarily free) paths P, P'
(one-dimensional submanifolds of $\mathcal{M}$) and carry arbitrary (but monotonic) clocks
τ, τ', exchange radar or light signals. Then, according to Fig. 2.1, the events e in an
appropriate neighborhood of parts of P, P' are characterized by the four numbers
$\tau_1, \tau_2, \tau_3, \tau_4$ (called radar coordinates) in a uniquely invertible manner.

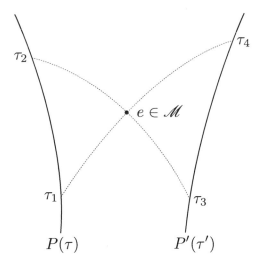

Fig. 2.1 The definition of radar coordinates $(\tau_1, \tau_2, \tau_3, \tau_4)$ on the basis of paths P and P' with clocks τ and τ'

Since there exist also quite strange and unphysical manifolds with locally Euclidean topology, one confines attention to paracompact and Hausdorff manifolds (see Abraham et al. 1988 for details). An important consequence of this is that there then exists a decomposition of unity in $\mathcal{M}$ which enables one to carry out integration on $\mathcal{M}$.

Manifolds $\mathcal{M}$ of this type automatically carry a vector and tensor structure, where the contravariant vectors $\mathbf{u}$ at an event $e \in \mathcal{M}$ are defined as the tangent vectors to differentiable curves through e. The dual space of linear functions $\underline{\sigma}(\mathbf{u}) \equiv \langle \underline{\sigma}, \mathbf{u} \rangle$ represents the covariant vectors or 1-forms. It is, however, important to stress that up to now these vector spaces are purely local, and the vector spaces at different events e_1 and e_2 are unrelated to each other. Tensors of type (r, s) in e are defined in the usual way as multilinear functions of s contravariant and r covariant vectors. Further important quantities, which automatically live in such a differential topological manifold, are the differential $\underline{df}$, the exterior derivative, and the Lie derivative:

- $\underline{df}$ is a 1-form associated with any differentiable function f on $\mathcal{M}$: $\langle \underline{df}, \mathbf{u} \rangle$ is the derivative of f at e along a curve with tangent vector $\mathbf{u}$.
- The exterior derivative is a generalization of the differential to so-called q-forms $\mathbf{A}$ which are antisymmetric tensors of type $(0, q)$. In coordinates, and with the symbol $\wedge$ for the antisymmetric product:

$$\mathbf{A} = A_{[\mu_1 \dots \mu_q]} \wedge \underline{dx}^{\mu_1} \wedge \dots \wedge \underline{dx}^{\mu_q}, \quad d\mathbf{A} = \underline{dA}_{[\mu_1 \dots \mu_q]} \wedge \underline{dx}^{\mu_1} \wedge \dots \wedge \underline{dx}^{\mu_q},$$

where the construction is, due to the antisymmetrization, coordinate independent. Applying the exterior derivative two times always gives zero, viz., $d(d\mathbf{A}) \equiv 0$, because the second derivatives of a two times continuously differentiable function are independent of the order. Geometrically, the relation $d(d\mathbf{A}) \equiv 0$

expresses the fact that the boundary of a manifold possesses no boundary. (We come back to this fact in connection with Einstein's field equations in Sect. 3.2.) The q-forms also provide a generalization of the Gauss and Stokes theorems in the form $\int_{\partial \mathscr{G}} \mathbf{A} = \int_{\mathscr{G}} d\mathbf{A}$ for any $(q+1)$-dimensional submanifold $\mathscr{G}$ of $\mathscr{M}$ with boundary $\partial \mathscr{G}$.

• The Lie derivative is based on the fundamental theorem of the theory of differential equations according to which any continuously differentiable vector field $\mathbf{u}(p)$ on $\mathscr{M}$ induces a congruence of integration curves $p_i(\lambda)$, which in turn induce a one-parameter group of diffeomorphisms Φ_λ. These diffeomorphisms Φ_λ also transport all vectors and tensors from an event p to the image point $\Phi_\lambda(p) = q$, and the Lie derivative of a tensor field $\mathbf{T}(p)$ is defined by

$$L_{\mathbf{u}} T = \lim_{\lambda \to 0} \frac{1}{\lambda} \{ \mathbf{T}(q) - \mathbf{T}(p \to q) \} .$$

For scalar fields the Lie derivative is obviously the same as the standard derivative. Of special interest is the Lie derivative of a vector field, viz., $L_{\mathbf{u}} \mathbf{v} = [\mathbf{u}, \mathbf{v}]$, which is identical with the commutator of two vector fields, and the vector fields with this composition form a Lie algebra with Jacobi identity.

For more details about these interesting quantities and operations in a topological manifold, see Abraham et al. (1988). It should be stressed, however, that these structures do not yet allow the formulation of physical field equations (differential equations) because the latter rest on the concept of partial derivatives of general fields at a point (event), whereas the exterior derivative is only defined for q-forms, and the Lie derivative also depends on derivatives at neighboring points. For the formulation of field equations, we therefore need additional structure on $\mathscr{M}$, at least an affine structure, something which results in a physically natural way from the conformal light cone structure (see below) and the projective inertial structure (see Sect. 2.3).

The phenomenon of light propagation in vacuum—with its independence of frequency, polarization, intensity, and velocity of source and receiver—endows the differential-topological manifold $\mathscr{M}$ with a new, characteristic structure: events in $\mathscr{M}$ are locally classified by the property of whether they can or cannot be connected by direct light signals. A systematic, operational, and fairly complete analysis of this structure (solely by coordinate-independent topological and incidence properties, and without further underlying structures such as a metric) was first presented in the seminal paper (Ehlers et al. 1972), which uses earlier partial results by R. Sachs, F.A.E. Pirani, A. Schild, A. Trautman, and R. Penrose, some of which are quoted in Ehlers et al. (1972). [An alternative, somewhat earlier, but less complete attempt to build up (pseudo-)Riemannian spacetime from hypotheses about the paths of light rays and free particles was provided in Castagnino (1971).]

According to experience the conformal structure obeys the following axioms:

Axiom L_0 *Light rays are smooth C^3-submanifolds of $\mathscr{M}$, diffeomorphic to $\mathbb{R}^1$.*

Fig. 2.2 The definition of the function $g(p) = \tau(e_1)\tau(e_2)$ in neighborhoods $\mathscr{U} \subset \mathscr{V}$ of the event e, and relative to a path P with clock τ

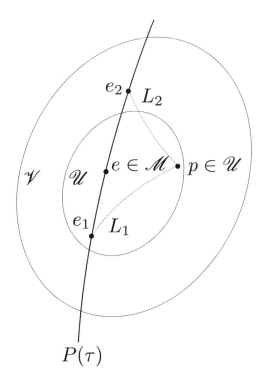

However, more than for particle paths, this axiom implies a gross extrapolation of real experimental facts because, due to the infinite time dilatation, a light ray cannot carry with it any instrument continuously measuring a path parameter. Strictly speaking, only the acts of emission and absorption of a photon can be registered.

Axiom L_1 *For any event $e \in \mathscr{M}$ and any (not necessarily free) particle P through e, there exist neighborhoods $\mathscr{V}$ and $\mathscr{U} \subset \mathscr{V}$ of e such that any event $p \in \mathscr{U}$, not lying on P, has precisely two light connections L_1, L_2 with P, lying completely in $\mathscr{V}$, and these meet P in distinct events e_1, e_2 (see Fig. 2.2). For an appropriate monotone parameter τ on $P \cap \mathscr{V}$ with $\tau(e) = 0$, the function $g(p) = \tau(e_1)\tau(e_2)$ is at least two times continuously differentiable in $\mathscr{U}$.*

The two light connections represent the decisive contrast with Newtonian space-time where the concept of absolute time forms the substitute for the whole conformal structure (compare Sect. 1.5). It should also be stressed here that, in strong gravitational fields, Axiom L_1 is no longer valid globally: effects like the light deflection by big masses can lead to more than two light connections between p and P, and so-called horizons can have the effect that only one light connection exists, or none at all. Axioms L_1 and L_2 already contain a part of Einstein's equivalence principle, so essential to general relativity: in gravitational fields, light has locally the same behavior as in special relativity.

Axiom L₂ *At an arbitrary event p, the manifold of vectors L_p in the direction of light rays separates the residual manifold D_p/L_p of direction vectors into two connected components, the timelike and spacelike vectors. The manifold of nonzero vectors in p, adjacent to light rays, consists of two connected components, the forward and backward light cones.*

This does not yet imply any intrinsic difference between future and past. However, due to the continuity of the light cone structure, it implies a local time orientation.

Some important and characteristic properties of the above function $g(p)$, which are independent of the specific particle path P and of the specific parameter τ, follow from the Axioms L_1 and L_2:

- $g(p) = 0$ for $p \in \mathcal{U}$ holds if and only if p lies on a light ray through e.
- In relation to an arbitrary coordinate system $\langle x^\mu \rangle$ in the neighborhood of e, we have $\partial g(p)/\partial x^\mu|_e \equiv g_{,\mu}(e) = 0$, for $\mu = 0, \ldots, 3$. If this were not the case, the equation $g(p) = 0$ would describe a smooth hyperplane with normal vector $g_{,\mu}(e)$, containing locally all light rays through e, and this would contradict Axiom L_2.
- For a general function $f(p)$, the second derivatives $f_{,\mu\nu}(e)$ do not constitute a tensor with correct transformation properties for general (non-linear) coordinate transformations. However, due to property (2), the derivatives $g_{,\mu\nu}(e)$ constitute a tensor.
- On a light ray through e, we have $g(x^\mu(\sigma)) \equiv 0$. Differentiating this equation twice with respect to σ leads to

$$g_{\mu\nu}k^\mu k^\nu = 0 , \tag{2.1}$$

with $g_{\mu\nu} = g_{,\mu\nu}(e)$, where $k^\mu = dx^\mu/d\sigma$ is an arbitrary lightlike vector at e. In principle, (2.1) can be used to calculate (up to a factor) the tensor $g_{\mu\nu}$ for a given coordinate system $\langle x^\mu \rangle$ and a given parametrization τ, i.e., by solving it for nine light rays with linearly independent products $k^\mu k^\nu$. If further light rays also solve (2.1), this constitutes a test of Axiom L_2.

- By linear coordinate transformations, the symmetric matrix $g_{\mu\nu}$ can be diagonalized, with diagonal elements $g_{\mu\mu} = 1, -1$, or 0. But only the signature $(1, -1, -1, -1)$ and its negative satisfy Axiom L_2. (In all other cases the nonzero lightlike vectors at e form only one connected component, or the non-lightlike vectors form only one connected component, contradicting Axiom L_2.) Therefore, in appropriate coordinates, $g_{\mu\nu}$ can be set locally in the form $g_{\mu\nu}(e) = \eta_{\mu\nu}$, known from special relativity.

Now, the tensor $g_{\mu\nu}$ can be used to define a *scalar product* (uniquely up to a positive factor) for vectors **u** and **v** at e :

$$g(\mathbf{u}, \mathbf{v}) = g_{\mu\nu}u^\mu v^\nu ,$$

and this product is invariant under all coordinate transformations. Two vectors $\mathbf{u}$ and $\mathbf{v}$ are *orthogonal* if $g(\mathbf{u}, \mathbf{v}) = 0$. Vectors $\mathbf{u}$ are *timelike* if $g(\mathbf{u}, \mathbf{u}) > 0$, *spacelike* if $g(\mathbf{u}, \mathbf{u}) < 0$, and *lightlike* if $g(\mathbf{u}, \mathbf{u}) = 0$. If two spacelike vectors $\mathbf{u}$ and $\mathbf{v}$ are orthogonal to one and the same timelike vector, their normed scalar product

$$\gamma(\mathbf{u}, \mathbf{v}) = \frac{g(\mathbf{u}, \mathbf{v})}{\sqrt{g(\mathbf{u}, \mathbf{u})g(\mathbf{v}, \mathbf{v})}} ,$$

ranges between -1 and $+1$, and can be interpreted as the cosine of the angle between $\mathbf{u}$ and $\mathbf{v}$. For two timelike vectors, we have $|\gamma(\mathbf{u}, \mathbf{v})| \geq 1$, and the quantity $\sqrt{1 - \gamma^{-2}}$ denotes the momentary relative velocity (in units of the light velocity c) of two particles with tangent vectors $\mathbf{u}$ and $\mathbf{v}$ at e.

Although the vectors $\mathbf{u}, \mathbf{v}, \ldots$ are, to begin with, only defined locally at the event e, the conformal structure is not confined to the local characterization of light cones at events $e, p, q, \ldots$. In fact, it also provides laws for the connection of light cones at neighboring events, and in this way leads to a differential equation for light rays. Indeed, one can show not only that $g_{\mu\nu}(e)$ can be transformed to the normal form $\eta_{\mu\nu}$, but that, by a coordinate transformation

$$x^\mu = x'^\mu - \frac{1}{2} K^\mu_{\nu\lambda}(e) x'^\nu x'^\lambda ,$$

with

$$K^\mu_{\nu\lambda}(e) = \frac{1}{2} g^{\mu\rho}(e) \left[g_{\rho\nu,\lambda}(e) + g_{\rho\lambda,\nu}(e) - g_{\nu\lambda,\rho}(e) \right] , \qquad (2.2)$$

the condition $g'_{\mu\nu,\lambda}(e) = 0$ can also be arranged for all λ, μ, ν, where $g^{\mu\nu}$ is the dual tensor to $g_{\mu\nu}$, defined by $g^{\mu\nu} g_{\nu\lambda} = \delta^\mu_\lambda$. (The property $g'_{\mu\nu,\lambda} = 0$ can even be achieved along a finite curve through e. It is important to stress that the functions $g_{\mu\nu,\lambda}$ cannot in general be understood as the third derivatives of the function $g(p)$ because $g_{\mu\nu}(e)$ and $g_{\mu\nu}(p)$ are defined differently at neighboring points.) Geometrically, the property $g'_{\mu\nu,\lambda}(e) = 0$ expresses the fact that the light cones v_p at neighboring events p have the same orientation and the same opening angles as v_e, i.e., if p lies on v_e, the light cone v_p touches v_e from the interior, as shown in Fig. 2.3. (Otherwise the point p would have three light connections with a particle P through e, contradicting Axiom L_1.)

The light ray from e to p obviously lies locally on both light cones and is therefore their unique contact line. In the coordinates $\langle x'^\mu \rangle$, this line is straight (and not a screw line!), and with an appropriate parameter τ, it is represented by $d^2 x'^\mu / d\tau^2 = 0$. Transforming back to the coordinates $\langle x^\mu \rangle$ and to a general parameter σ leads to the *general equation for a conformal null geodesic*, viz.,

$$\frac{dk^\mu}{ds} + K^\mu_{\nu\lambda} k^\nu k^\lambda = C k^\mu , \qquad (2.3)$$

Fig. 2.3 If the point p lies on the light cone ν_e, the light cone ν_p touches ν_e from the interior

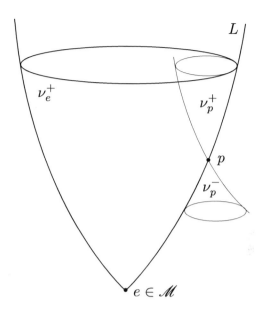

with $C = -(d^2s/d\tau^2)/(ds/d\tau)$. The quantities $K^{\mu}_{\nu\lambda}$ are called the *conformal connections*.

Although the coefficients $K^{\mu}_{\nu\lambda}$ do not constitute a tensor, they can be used to define a *covariant differentiation* of vector and tensor fields: whereas the derivatives $\partial S^{\nu}/\partial x^{\mu}$ of a vector field behave as tensor components only for linear coordinate transformations, the quantities

$$\tilde{\nabla}_{\mu} S^{\nu} = \frac{\partial S^{\nu}}{\partial x^{\mu}} + K^{\nu}_{\mu\lambda} S^{\lambda}$$

do this for general transformations. A coordinate-independent vector transport (a kind of parallel transport) can then be defined along a curve $x^{\mu}(\tau)$ with tangent vector $T^{\mu} = dx^{\mu}/d\tau$, viz.,

$$\frac{\tilde{D} S^{\nu}}{D\tau} = T^{\mu} \left(\frac{\partial S^{\nu}}{\partial x^{\mu}} + K^{\nu}_{\mu\lambda} S^{\lambda} \right) = 0 . \tag{2.4}$$

However, this transport is not conformally invariant. A gauge transformation

$$g_{\mu\nu}(x^{\lambda}) \rightarrow \Omega^2(x^{\lambda}) g_{\mu\nu}(x^{\lambda})$$

results in

$$K^{\mu}_{\nu\lambda} \longrightarrow \overline{K}^{\mu}_{\nu\lambda} = K^{\mu}_{\nu\lambda} + \frac{1}{2\Omega^2} \left(\delta^{\mu}_{\nu} \Omega^2_{,\lambda} + \delta^{\mu}_{\lambda} \Omega^2_{,\nu} - g_{\nu\lambda} \Omega^{2,\mu} \right) , \tag{2.5}$$

whereby just the trace terms of $K^\mu_{\nu\lambda}$ change. [In a given coordinate system, the coefficients $K^\mu_{\nu\lambda}(e)$ can all be made zero by an appropriate gauge transformation.]

Even though the conformal structure alone does not provide unique directional derivatives of vectors and tensors, and therefore no covariant differential equations, it already allows the introduction of mathematical concepts that are central to the full theory of general relativity. For instance, a covariant and gauge-invariant transport can be defined for surface elements containing the tangent vector T^μ (of unit length), or for vector fields $S^\nu(\tau)$ orthogonal to T^μ, and therefore for orthogonal vierbeins accompanying an arbitrary non-lightlike curve, the so-called *rotation-free* or *Fermi transport* (Fermi 1922):

$$\frac{dS^\mu}{d\tau} = -K^\mu_{\nu\lambda}T^\nu S^\lambda + \left(S_\nu \frac{DT^\nu}{D\tau}\right)T^\mu . \tag{2.6}$$

According to Synge (1966, Sect. III.8), the Fermi transport along timelike curves can be visualized by a nice thought experiment with so-called bouncing photons. An observer carries along his path $x^\mu(\tau)$ a telescope-like instrument, with which he can emit and receive light rays. If he adjusts the direction S^ν (orthogonal to T^μ) of the instrument in such a way that light reflected on nearby objects is always exactly focused back into the telescope, the instrument is Fermi transported. [A simple proof of this fact was given in Pirani (1965).]

As we have seen, the Axioms L_1 and L_2 imply that the light cone structure $g_{\mu\nu}(x)$ is locally identical with the light cone structure of Minkowski spacetime in a neighbourhood of first order, i.e., for all derivatives $g_{\mu\nu,\lambda}(e)$. However, this should not be expected to continue to, e.g., the second derivatives $g_{\mu\nu,\lambda\kappa}(e)$. Rather, these quantities will signify some characteristic and invariant physical properties of the global light cone structure, and of the manifold $\mathcal{M}$ itself, which may, for instance, be caused by strong gravitational fields. There are at least three ways to construct some light-geometric quantities from the non-tensorial derivatives $g_{\mu\nu,\lambda\kappa}(e)$. Purely mathematically, one can try to construct a gauge-invariant tensor from these derivatives. More geometrically, one can ask whether the above Fermi transport induces a holonomy group, i.e., a non-trivial map of vectors due to transport around a closed curve. Here, we shall follow a third route, namely the possible change in shape of a thin bunch of parallel light rays.

The 'central' ray $x^\mu(\tau)$ obeys (in distinguished parametrization) the null geodesic equation

$$\frac{d^2 x^\mu}{d\tau^2} + K^\mu_{\nu\lambda}\frac{dx^\nu}{d\tau}\frac{dx^\lambda}{d\tau} = 0 .$$

The neighboring light rays $x^\mu(\tau) + \delta x^\mu(\tau)$ obey equivalent equations. Subtracting the two equations, linearization in the small quantities $\delta x^\mu(\tau)$, and 'covariantiza-tion' leads to

$$\frac{\tilde{D}}{D\tau}\delta x^\mu = \frac{d}{d\tau}\delta x^\mu + K^\mu_{\nu\lambda}k^\nu \delta x^\lambda . \tag{2.7}$$

Covariant differentiation of this equation results together with the geodesic equation and the equation for $d^2(\delta x^\nu)/d\tau^2$ in the *equation of geodesic deviation*:

$$\frac{\tilde{D}^2}{D\tau^2}\delta x^\mu = R^\mu_{\nu\lambda\kappa}k^\nu k^\kappa \delta x^\lambda \; , \tag{2.8}$$

with

$$R^\mu_{\nu\lambda\kappa} = K^\mu_{\nu\kappa,\lambda} - K^\mu_{\nu\lambda,\kappa} + K^\mu_{\lambda\rho}K^\rho_{\nu\kappa} - K^\mu_{\kappa\rho}K^\rho_{\nu\lambda} \; .$$

Since, according to (2.8), $R^\mu_{\nu\lambda\kappa}$ maps the vector triple $k^\nu, k^\kappa, \delta x^\lambda$ to the vector $\tilde{D}^2(\delta x^\mu)/D\tau^2$, it is to be expected, according to the quotient theorem, that $R^\mu_{\nu\lambda\kappa}$ will be a type $(1,3)$ tensor, and this is indeed the case.

It is easily checked that the tensor $R_{\mu\nu\lambda\kappa} = g_{\mu\rho}R^\rho_{\nu\lambda\kappa}$ has many interesting symmetries, most of which also have a geometrical meaning. Antisymmetry in the first two indices and in the last two indices, and cyclic permutation symmetry

$$R_{\mu\nu\lambda\kappa} + R_{\mu\lambda\kappa\nu} + R_{\mu\kappa\nu\lambda} = 0 \; ,$$

from which follows also the pair symmetry $R_{\mu\nu\lambda\kappa} = R_{\lambda\kappa\mu\nu}$. These reduce the originally $4^4 = 256$ independent components to only 20. (This also follows from the following argument. There are 100 independent quantities $g_{\mu\nu,\lambda\kappa}$. In order that there should appear no third derivatives $\partial^3 x'^\mu/\partial x^\nu \partial x^\lambda \partial x^\kappa$ in coordinate transformations of the sought-for tensor, there must be 80 constraints.)

Since the vector transport of (2.7) is only covariant but not gauge invariant, the tensor $R^\mu_{\nu\lambda\kappa}$ will not be gauge invariant either. However, since we are dealing with a tensor, it is relatively easy to extract a gauge invariant tensor from it, in contrast to the non-tensorial quantities $K^\mu_{\nu\lambda}$. Equation (2.5) shows that gauge transformations change the trace terms of $K^\mu_{\nu\lambda}$, and therefore also those of $R^\mu_{\nu\lambda\kappa}$. So we just have to subtract the traces of $R^\mu_{\nu\lambda\kappa}$, and this under preservation of the symmetries. Due to these symmetries, the tensor $R^\mu_{\nu\lambda\kappa}$ has only one independent simple trace $R_{\nu\kappa} = R^\mu_{\nu\mu\kappa}$, and one double trace $R = g^{\nu\kappa}R_{\nu\kappa}$. Subtraction of these traces leads uniquely to the *Weyl tensor* or *conformal curvature tensor*

$$C^\mu_{\nu\lambda\kappa} = R^\mu_{\nu\lambda\kappa} - \frac{1}{2}\left(\delta^\mu_\lambda R_{\nu\kappa} - \delta^\mu_\kappa R_{\nu\lambda} - g_{\nu\lambda}R^\mu_\kappa + g_{\nu\kappa}R^\mu_\lambda\right) + \frac{1}{6}\left(\delta^\mu_\lambda g_{\nu\kappa} - \delta^\mu_\kappa g_{\nu\lambda}\right)R \; , \tag{2.9}$$

which H. Weyl derived in Weyl (1918a), as mentioned in Sect. 2.1. [For a geometric construction of Weyl's conformal curvature tensor via light rays, see Pirani and Schild (1966)]. Due to its trace-freeness, the Weyl tensor has only ten independent components.

Coming back to the change in shape of a parallel light bundle from which we started, we use a basis $(\mathbf{e}_0, \mathbf{e}_1, \mathbf{e}_2, \mathbf{e}_3)$, where $\mathbf{e}_0 = \mathbf{k}$, while $\mathbf{e}_1$ is the timelike vector tangent to the path $u^\mu(\tau)$ of an observer measuring the cross-section of the light bundle, and $\mathbf{e}_2$ and $\mathbf{e}_3$ are orthogonal to each other and orthogonal to $\mathbf{e}_0$ and $\mathbf{e}_1$. Then according to (2.8), the change in shape is determined solely by $S_\kappa^\mu = C_{0\kappa 0}^\mu$ with $\mu, \kappa = 2, 3$. The tensor S_κ^μ is trace-free (an expansion of the shape is not definable in a conformally invariant manner!), and due to the pair symmetry of the Weyl tensor, $S_{\mu\kappa}$ is symmetric (no rotation or vorticity is definable in a conformally invariant manner!), i.e., it has the form

$$S_{\mu\kappa} = \begin{pmatrix} a & b \\ b & -a \end{pmatrix} ,$$

with eigenvalues $\lambda_{1,2} = \pm\sqrt{a^2 + b^2}$. Therefore, a light bundle starting with spherical shape typically attains an elliptic shape. In relation to an observer with four-velocity u^μ, the Weyl tensor separates into an 'electric part' $E_{\mu\nu} = C_{\mu\nu\lambda\kappa}u^\lambda u^\kappa$ and a 'magnetic part' $H_{\mu\nu} = \varepsilon_{\mu\lambda\rho\sigma}C_{\nu\kappa}^{\rho\sigma}u^\lambda u^\kappa/2$, where $\varepsilon_{\mu\lambda\rho\sigma}$ is the totally antisymmetric tensor with $\varepsilon_{0123} = \sqrt{-\det(g_{\mu\nu})}$. In Newtonian gravity (with potential Φ), the equivalent to $E_{\mu\nu}$ is the 3-tensor $\Phi_{,ij} - \eta_{ij}\Phi_{,k}^{k}/3$, and the equivalent to $H_{\mu\nu}$ is identically zero.

As already mentioned in Sect. 2.1, the Weyl tensor completely determines the gravitational field in all vacuum parts of a physical system, and is particularly important for the study of gravitational waves. Whereas the spacetime variation of $C_{\nu\lambda\kappa}^\mu(x^\alpha)$ is quite versatile and indirectly (non-linearly and non-locally) determined by the matter parts of the physical system, there exists a quite simple and coordinate-independent classification of the local structure of $C_{\nu\lambda\kappa}^\mu(x^\alpha)$, the so-called *Petrov classification*. It is based on the eigenvalue equation

$$(C_{\mu\nu\lambda\kappa} - \lambda g_{\mu\nu\lambda\kappa})F^{\lambda\kappa} = 0 ,$$

where the 'unity tensor' has the form $g_{\mu\nu\lambda\kappa} = g_{\mu\lambda}g_{\nu\kappa} - g_{\mu\kappa}g_{\nu\lambda}$, thanks to the symmetry properties. Due to the antisymmetry of $C_{\mu\nu\lambda\kappa}$ in the first two and in the last two indices, this eigenvalue equation is best studied on the six-dimensional space of antisymmetric bivectors

$$F^{[\lambda\kappa]} = F^A , \quad \text{with } A = \begin{pmatrix} 1 & 2 & 3 & 4 & 5 & 6 \\ (01) & (02) & (03) & (23) & (31) & (12) \end{pmatrix} ,$$

where it takes the form

$$(C_{AB} - \lambda g_{AB})F^A = 0 , \tag{2.10}$$

and in the local inertial system ($g_{\mu\nu} = \eta_{\mu\nu}$) the tensor C_{AB} has the form

$$\begin{pmatrix} M & N \\ N & -M \end{pmatrix},$$

with symmetrical and trace-free 3×3 matrices M and N, and with

$$g_{AB} = \begin{pmatrix} -1 & 0 \\ 0 & +1 \end{pmatrix}.$$

We then have, for the eigenvalues λ, two complex conjugated equations of degree three with solutions $\lambda_{1/4} = \alpha_1 \pm i\beta_1, \lambda_{2/5} = \alpha_2 \pm i\beta_2, \lambda_{3/6} = -\lambda_{1/4} - \lambda_{2/5}$. The most general Petrov type I is realized for $\lambda_1 \neq \lambda_2 \neq \lambda_3 \neq \lambda_1$. The case $\lambda_1 \neq \lambda_2 = \lambda_3$ is Petrov type II. A special subcase of type II (with simpler normal forms of the matrices M, N) is called type D. The Schwarzschild and Kerr solutions even belong globally to this type. The case $\lambda_1 = \lambda_2 = \lambda_3 = 0$ is Petrov type III, with a special subcase of type N. [For more details about this Petrov classification, see Stephani et al. (2003, Chap. 4). There one also finds a characterization of the different Petrov types by the so-called principal null directions. A nice characterization of the different Petrov types by Gedanken experiments with a so-called gravitational compass was provided in Szekeres (1965).] An especially useful application of the Petrov classification shows up in the *peeling theorem* of R. Sachs in Sachs (1961), concerning the asymptotic ($r \to \infty$) behavior of a vacuum gravitational field far from any material sources. In an expansion in powers of r^{-1}, the Weyl tensor then has the symbolic form

$$C_{\mu\nu\lambda\kappa} = \frac{N}{r} + \frac{III}{r^2} + \frac{II}{r^3} + \frac{I}{r^4}. \tag{2.11}$$

We have already mentioned that the Kerr metric is of Petrov type D. Indeed, this important solution of Einstein's vacuum field equations was found by R. Kerr in Kerr (1963), not so much as a realization of a physical target, but by lucky circumstances in a mathematical study of Petrov type D solutions. And whereas it was originally interpreted as the exterior solution of a real astrophysical spinning mass (with the aim of finding a corresponding interior solution), it was only proven much later that this solution applies uniquely to all rotating black holes (Heusler 1996). (Some authors still argue incorrectly that the Kerr metric also represents the exterior of real, non-collapsed rotating astrophysical bodies.) As far as we know, there is still no simple and mainly physically motivated derivation of the Kerr metric. However, there does exist a derivation via the solution of a boundary value problem (at the horizon) using the (mathematically involved) inverse scattering method (Meinel 2013).

2.3 Free Particles Define the Projective Structure (Desargues Theorem)

As already mentioned in the previous section, the conformal structure of spacetime alone does not yet suffice to formulate differential equations for physical fields. (Moreover, the conformal or scale invariance of the light structure cannot extend to all other physical laws and effects in the material world, e.g., a 100 times bigger Eiffel tower on a 100 times bigger Earth would surely collapse.) If we look at other elementary and universal one-dimensional paths in the physical spacetime manifold $\mathscr{M}$, by analogy with the light rays of Sect. 2.2, the only distinguished class of such objects are the so-called free particles, as defined in Sect. 1.3. And indeed, these objects do seem to endow spacetime with a further fundamental structure, namely a unique inertial or projective structure, because modern experiments have confirmed the universality of free fall (part of Einstein's equivalence principle) to an accuracy of 1.8×10^{-13} (Schlamminger et al. 2008), while an improvement up to 10^{-18} seems possible with planned satellite experiments [see Sect. 1.5 and the papers 184001–184013 in the Focus issue *Tests of the weak equivalence principle* in Class. Quant. Grav. **29/18** (September 2012)]. This suggests, in agreement with Ehlers, Pirani, Schild (EPS), the following axiom.

Axiom P$_1$ *For any event $e \in \mathscr{M}$, and for any timelike direction (with respect to the conformal structure) at e, there exists one and only one free particle path P through e with this direction.*
As Ehlers noted in Ehlers (1985, p. 87):

> The projective structure represents that unit of inertia and gravity which Einstein substituted for both, and for which Weyl introduced the suggestive term *guiding field* (*Führungfeld*).

In a next step, we have to characterize by geometric and/or mathematical means the distinguished inertial structure which these free particle paths inscribe within the manifold $\mathscr{M}$. By analogy with the conformal structure, as is evident in particular from Axiom L$_1$, it is to be expected that in general (e.g., in general and strong gravitational fields) such a characterization can only be given locally in a neighborhood of the event e. EPS do this in their Axiom P$_2$ only in a coordinate-dependent way and not by 'qualitative (incidence and differential–topological) properties' as they announce in their introduction. One can criticize this approach in the same way as Einstein (as already quoted in Sect. 1.4): "Who does not feel the painfulness of such a formulation?" In somewhat later publications of Pirani (1973) and Ehlers and Schild (1973), there were more geometrical characterizations of the structure of free-fall paths (e.g., a reconstruction of the projective curvature tensor from particle paths) by so-called tongs and comb constructions, or by a zig-zag construction. And still later, there appeared characterizations in Ehlers and Köhler (1977) and Coleman and Korte (1980) in terms of a maximal local isotropy or dilatation symmetry.

However, we would like here to advocate a simpler and more elementary characterization of this structure, using only incidence properties between the free-fall paths. In Sect. 1.4, we succeeded in providing such an alternative characterization of the free particle paths (the straight lines in the arena of flat spacetime of classical mechanics and special relativity) by exploiting the geometric incidence figure of Desargues (see Fig. 1.2 in Sect. 1.4), and with the help of a proof by Hilbert. This raises the question as to whether this recipe can be extended to the curved manifold of general relativity via a 'localization of the Desargues theorem', and of course without the parallel axiom used by Hilbert in his proof. In Heilig and Pfister (1990), it was shown that this is indeed possible.

Axiom P$_2$ *A path structure is a free fall (or inertial) structure if and only if, for any point e of the manifold $\mathcal{M}$, an ε-neighbourhood can be found such that the path structure obeys there the Desargues incidence properties up to order ε^2.*

In the following paragraph and in Appendix A, we summarize the main steps of the proof given in Heilig and Pfister (1990) that this Axiom P$_2$ is indeed equivalent to the coordinate-dependent Axiom P$_2$ of EPS.

The 'only if' part of Axiom P$_2$, i.e., that the free-fall paths fulfil the Desargues properties up to order ε^2, is almost trivially evident. In the local inertial system, all free-fall paths in the ε-neighbourhood of e have the representation

$$x^\mu(\tau) = x^\mu(e) + \tau v^\mu(e) + O(\varepsilon^3) \,,$$

with $v^\mu(e) = \mathrm{d}x^\mu/\mathrm{d}\tau(e)$, and where the parameter τ is of order ε. That these straight lines (up to errors of order ε^3) fulfil the Desargues properties, is a standard result of projective geometry, as explicated in Sect. 1.4.

In contrast, the proof of the converse, that a manifold of paths locally satisfying the Desargues properties is necessarily a free-fall path structure (consisting of projective geodesics), is quite involved. Here we only give a qualitative indication of the strategy leading to this result. We begin with the special case shown in Fig. 1.2 of Sect. 1.4, where the directions of the paths P, P', P'' at e are linearly dependent. The existence of the intersection points e_1, e_2, e_3 to order ε^2, due to the local Desargues theorem, guarantees that in this case the whole Desargues configuration can be reduced to a two-dimensional submanifold. [A detailed proof is to be found in Heilig and Pfister (1990, Sect. 2.2).] This is the so-called surface-forming property of the projective geodesics that also played a major role in Pirani (1973) and Ehlers and Schild (1973).

As usual in projective geometry, the situation in two dimensions is special, and the most difficult to prove. From the Desargues properties there follows an involved functional equation for the central 'acceleration function' $\mathbf{K}(\mathbf{v})$ of the path structure, defined by the local expansion of the particle path

$$\mathbf{x}(\tau) = \mathbf{x}(0) + \tau\mathbf{v}(0) + \tau^2\mathbf{a}(\mathbf{x}, \mathbf{v}) + O(\varepsilon^3) \approx \mathbf{x}(0) + \tau\mathbf{v}(0) + \tau^2\mathbf{a}(0, \mathbf{v}) + O(\varepsilon^3) \,,$$

with $\mathbf{a}(0, \mathbf{v}) =: \mathbf{K}(\mathbf{v})$, and it can be shown that the only admissible solutions of this functional equation are such that $\mathbf{K}(\mathbf{v})$ depends linearly and symmetrically bilinearly on the components of the two-dimensional direction vector $\mathbf{v}$. However, such dependencies can be eliminated by suitable coordinate and parameter transformations, i.e., $\mathbf{K}(\mathbf{v})$ can be made identically zero. The generalization to higher-dimensional cases consists then in a simple and more or less formal extension of the two-dimensional results. Some details of the (technically quite involved) proof for the elimination of the acceleration function $\mathbf{K}(\mathbf{v})$ are deferred to Appendix A. Even more details can of course be found in Heilig and Pfister (1990).

According to Axioms P_1 and P_2, and due to the proof that the local Desargues property is equivalent to the coordinate-dependent EPS characterization of free-fall paths, these have in the local inertial system the representation $d^2 x^\mu / d\tau^2(e) = 0$. In a different arbitrary coordinate system $\langle x'^\mu(x^\nu)\rangle$ and with a new parameter $\sigma(\tau)$, they have the form

$$\frac{d^2 x'^\mu}{d\sigma^2} + \Pi_{\nu\lambda}^\mu(x'^\alpha) \frac{dx'^\nu}{d\sigma} \frac{dx'^\lambda}{d\sigma} = k(\sigma) \frac{dx'^\mu}{d\sigma} , \qquad (2.12)$$

with the factor $k(\sigma) = (d\sigma/d\tau)(d^2\tau/d\sigma^2)$, and with the symmetrical *projective connections*

$$\Pi_{\nu\lambda}^\mu(x'^\alpha) = -\frac{\partial^2 x'^\mu}{\partial x^\rho \partial x^\sigma} \frac{\partial x^\rho}{\partial x'^\nu} \frac{\partial x^\sigma}{\partial x'^\lambda} . \qquad (2.13)$$

Paths of the form (2.12) are called *projective geodesics*, and this in all cases, whether their tangent vectors $dx'^\nu/d\sigma$ are timelike, lightlike, or spacelike. In analogy with the conformal structure in Sect. 2.2, the projective structure induces in $\mathcal{M}$ a directional derivative, the so-called *covariant derivative* for arbitrary tensor fields:

$$T_{\nu_1\dots\nu_s;\lambda}^{\mu_1\dots\mu_r} = \frac{\partial}{\partial x^\lambda} T_{\nu_1\dots\nu_s}^{\mu_1\dots\mu_r} + \sum_{n=1}^{r} \Pi_{\lambda\rho}^{\mu_n} T_{\nu_1\dots\nu_s}^{\mu_1\dots\rho\dots\mu_r} - \sum_{m=1}^{s} \Pi_{\lambda\nu_m}^{\rho} T_{\nu_1\dots\rho\dots\nu_s}^{\mu_1\dots\mu_r} , \qquad (2.14)$$

and this operation is independent of the chosen coordinate system. The exterior and Lie derivatives introduced in Sect. 2.2 are identical with the following covariant derivatives:

$$d\mathbf{A} = \mathbf{A}_{\mu_1\dots\mu_q;\lambda} d\mathbf{x}^\lambda \wedge d\mathbf{x}^{\mu_1} \wedge \dots \wedge d\mathbf{x}^{\mu_q} ,$$

and

$$(L_\mathbf{X}\mathbf{Y})^\mu = Y^\mu{}_{;\nu} X^\nu - X^\mu{}_{;\nu} Y^\nu .$$

The covariant derivative induces *Levi-Civita parallel transport* (Levi-Civita 1917) along a curve $x^\mu(\tau)$, e.g., for a contravariant vector Y^μ:

$$\frac{DY^\mu}{D\tau} = \frac{dY^\mu}{d\tau} + \Pi_{\nu\lambda}^\mu Y^\nu \frac{dx^\lambda}{d\tau} = 0 . \qquad (2.15)$$

A free particle path or projective geodesic is then an auto-parallel. However, the above parallel transport is not invariant under parameter transformations (projective transformations) of the curve $x^\mu(\tau)$. Rather, in transformations $\tau \rightarrow \sigma$, the projective connections transform to

$$\Pi'^\mu_{\nu\lambda} = \Pi^\mu_{\nu\lambda} + (\delta^\mu_\nu q_\lambda + \delta^\mu_\lambda q_\nu) , \quad \text{with} \quad q_\lambda \frac{dx^\lambda}{d\sigma} = -\frac{1}{2}k(\sigma) .$$

Therefore, this transport is not really a transport of direction vectors but, similarly to the Fermi transport in Sect. 2.2, a transport of bivectors or surface elements. And once again, for timelike curves $x^\mu(\tau)$, this transport can be visualized by the emission and reflection of free particles in the neighborhood of the observer path $x^\mu(\tau)$, where, due to the surface-forming property of the family of geodesics, this construction is independent of the velocity of the emitted particles (see Pirani 1973; Ehlers and Schild 1973).

By analogy with the Fermi transport, one can ask whether this parallel transport induces a characteristic map $\mathbf{S} \rightarrow \mathbf{W}$ of a vector $\mathbf{S}$ after its parallel transport around an infinitesimal closed 'parallelogram' built up by vectors $\mathbf{T}$ and $\mathbf{U}$. Indeed, up to parametrization-dependent terms proportional to $\mathbf{S}, \mathbf{T}, \mathbf{U}$, we get

$$W^\mu = Q^\mu_{\nu\rho\sigma} S^\nu T^\rho U^\sigma = T^\rho U^\sigma \left(S^\mu_{;\sigma\rho} - S^\mu_{;\rho\sigma} \right) , \tag{2.16}$$

with

$$Q^\mu_{\nu\rho\sigma} = \Pi^\mu_{\nu\sigma,\rho} - \Pi^\mu_{\nu\rho,\sigma} + \Pi^\mu_{\rho\lambda}\Pi^\lambda_{\nu\sigma} - \Pi^\mu_{\sigma\lambda}\Pi^\lambda_{\nu\rho} .$$

[Compare with the expression for $R^\mu_{\nu\lambda\kappa}$ in the formula after (2.8) of Sect. 2.2.] Again this tensor has characteristic symmetry properties, although fewer symmetries than the tensor $R^\mu_{\nu\lambda\kappa}$. And since it is a tensor, we can get rid of the parameter-dependent parts by subtracting the traces $Q_{\mu\nu} = Q^\lambda_{\mu\lambda\nu}$, resulting in the *projective curvature tensor*

$$P^\mu_{\nu\rho\sigma} = Q^\mu_{\nu\rho\sigma} - \frac{1}{5}\delta^\mu_\nu(Q_{\rho\sigma} - Q_{\sigma\rho}) - \frac{1}{15}\delta^\mu_\rho(4Q_{\nu\sigma} + Q_{\sigma\nu}) + \frac{1}{15}\delta^\mu_\sigma(4Q_{\nu\rho} + Q_{\rho\nu}) , \tag{2.17}$$

which was also derived by Weyl (1921). In principle, a classification of this tensor (by analogy with the Petrov classification of the Weyl tensor) would be possible. However, due to its 64 independent components (in four dimensions), this would be much more involved, and it has not yet found any physically convincing interpretation. A manifold with $P^\mu_{\nu\rho\sigma} \equiv 0$ is said to be *projectively flat*. Due to the surface-forming property, a two-dimensional manifold is always projectively flat. In complete analogy with the significance of the Weyl tensor for the conformal structure, the tensor $P^\mu_{\nu\rho\sigma}$ contains all non-trivial information (going beyond special relativity) about the projective structure of $\mathcal{M}$.

The formulation of Newton's first law in Sect. 1.4 (in the absence of gravity) is of course no longer true in general gravitational fields, and in general there are no objects whatsoever which move globally on straight lines, because gravity cannot be shielded. (Any 'shielding device' would only bring in more gravity!) Thus, the term 'inertial system' for a system of straight axes in classical mechanics is (in the presence of gravity) somewhat misleading. Of course, one can in principle formulate any field of physics in any (global and unique) coordinate system one likes. That systems based on straight lines have nevertheless been successful (and even optimal) in Newton's *Principia* and in the modern teaching of mechanics and other fields of physics is due to 'lucky circumstances' in our Earth-based laboratories, and also in most (but not all!) other places in our universe, conditions which are usually kept secret in textbooks: the gravitational field on Earth, although not really weak (in comparison to other forces), is extremely homogeneous on any laboratory scale. Therefore, this 'gravitational force' can be nearly transformed away, e.g., by going to the famous Einstein elevator. The effects of Einstein's curvature of spacetime are by factors 10^{-9} smaller than the Newtonian gravitational force in any Earth-based laboratory or Earth satellite. [As the nice example with the tracks of balls and bullets in Misner et al. (1973, Box 1.6) shows, it is really the curvature of spacetime, and not of space!]

One may speculate that physics as we know it would never have been developed if the equivalence principle had been severely violated in nature, or if spacetime on our human scale had been grossly curved. In typical presentations of the laws of classical mechanics, gravity is treated like any other force field, in the arena of (hypothetical) global inertial systems. However, it was the great insight of Einstein, in his equivalence principle, that gravity is a very special 'force', and that gravity is intimately connected with inertia. According to Einstein's proposal, it is therefore advisable to use this unique structure provided by the paths of free particles under the common (and partly indistinguishable) action of inertia and gravity to build a new physical geometry of (pseudo-)Riemannian type. Then (2.12), with the projective coefficients $\Pi^{\mu}_{\nu\lambda}(x'^{\alpha})$ replaced by the (metric) Christoffel symbols $\Gamma^{\mu}_{\nu\lambda}(x'^{\alpha})$, established in the next section, continues to describe the global paths of free particles (geodesics) in arbitrary gravitational fields. And since all coefficients $\Gamma^{\mu}_{\nu\lambda}(x'^{\alpha})$ can simultaneously be made zero at a given event, one recovers a local version of Newton's first law in arbitrary gravitational fields:

Newton's First Law *In an ε-neighborhood around a given event, and in an appropriate class of coordinate systems, free particles (in the presence of gravity) move on straight lines up to errors of order ε^3.*

That these particles do not follow straight lines globally (and, in consequence, that free particles that emanate from a common event can be focused) is a result of 'real gravitational fields'. And these are mathematically encoded in the curvature of the (pseudo-)Riemannian spacetime, i.e., in the derivatives of the Christoffel symbols, which are, for their part, fixed (through Einstein's field equations) by the matter content in this region of spacetime (see Sect. 2.4). If one is unhappy with the above coordinate-dependent formulation of the local Newton law, it is very reassuring that a local version of the coordinate-independent characterization of straight lines through the Desargues property also exists, through Axiom P_2.

2.4 Weyl Structure, Affine Structure, and Metric Structure

So far the conformal structure (as established in Sect. 2.2) and the projective structure (in Sect. 2.3) have been introduced quite independently of each other. However, it turned out that the mathematical characteristics (connections, geodesics, vector transports, curvatures) of these structures were quite similar. Furthermore, physically, it is well known that the light velocity can be seen as the limit of the velocities of particles with nonzero mass. [For a long time this was optimally tested by high-energy electrons from accelerators. More recently the best test comes from astrophysics: neutrinos and light from the supernova SN1987A in the Large Magellanic Cloud reached the Earth at the same time to an accuracy of 2×10^{-9}, after a journey of 160,000 years (Longo 1987).] EPS formalize this experimental fact as a *compatibility property* between the conformal and projective structure.

Axiom C_1 *Each event e has a neighborhood $\mathcal{U}$ such that an event $p \in \mathcal{U}, p \neq e$ lies on a free particle P through e if and only if p is contained in the interior of the light cone v_e of e.*

From this, it follows immediately that a projective geodesic which represents a free particle cannot be spacelike anywhere (with respect to the conformal structure). One can also show that a projective geodesic whose tangent vector is somewhere lightlike, is a null geodesic. Referring to Fig. 2.4, we consider an event $p \in v_e$, and a

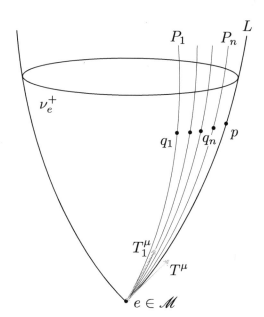

Fig. 2.4 Particle paths $P_1, \ldots, P_n$, approaching the light cone v_e. Their tangent vectors T_n^{μ} at e approach a lightlike vector T^{μ}, which is tangent to v_e

series of events q_n from the interior of v_e (and in an appropriate neighborhood $\mathcal{U}$ of e) with $\lim_{n\to\infty} q_n = p$. Then, Axiom C_1 states that there are free particles $P_n(\tau_n)$ connecting e and q_n. The parameters τ_n of these paths can be chosen in such a way that, in the geodesic equation (2.12) of Sect. 2.3, all k_n are zero, and also such that $P_n(\tau_n = 0) = e$ and $P_n(\tau_n = 1) = q_n$. Standard theorems of differential equations then imply that the series of tangent vectors T_n^μ at e converges: $\lim_{n\to\infty} T_n^\mu = T^\mu$, and T^μ can only be timelike or lightlike. If T^μ were timelike, it would define a free particle path P meeting the event $p \in v_e$, contradicting axiom C_1. Therefore we have $g_{\mu\nu} T^\mu T^\nu = 0$. If $\frac{\mathrm{d}}{\mathrm{d}\tau}(g_{\mu\nu} T^\mu T^\nu) \neq 0$, the quantity $g_{\mu\nu} T^\mu T^\nu$ would have to change sign in a neighborhood $\mathcal{U}$, and T^μ would have to become spacelike, contradicting the fact that P is the limit of free particle paths. Since P obeys the geodesic equation (of second order), it must therefore be lightlike throughout $\mathcal{U}$. If P entered the interior of v_e at some event q, there would be two paths between e and q, one being a free particle path with timelike tangent at e, the other the projective geodesic P with lightlike tangent at e. This contradicts the unique local definition of a free particle path by two events. Since P is then everywhere lightlike and lies completely in the null hyperspace v_e, a lightlike projective geodesic is necessarily a null geodesic.

It should be clear in this way that the Axiom C_1 also establishes a mathematical relation between the connections $K_{\nu\lambda}^\mu$ and $\Pi_{\nu\lambda}^\mu$ whose difference $\Delta_{\nu\lambda}^\mu = \Pi_{\nu\lambda}^\mu - K_{\nu\lambda}^\mu$ behaves as a type (1,2) tensor. For a conformal and projective geodesic with tangent vector T^μ, both (2.3) of Sect. 2.2 (with factor k_2) and (2.12) of Sect. 2.3 (with factor k_1) are valid. Subtraction of these equations results in $\Delta_{\nu\lambda}^\mu T^\nu T^\lambda = (k_1 - k_2) T^\mu$, and contraction of this equation with T_μ gives $\Delta_{\nu\lambda}^\mu T_\mu T^\nu T^\lambda = 0$ for all lightlike vectors T^μ. From this, and from the symmetry of $\Pi_{\nu\lambda}^\mu$ in the lower indices, it follows that

$$\Pi_{\nu\lambda}^\mu = K_{\nu\lambda}^\mu + \left(g_{\nu\lambda} b^\mu + \delta_\lambda^\mu c_\nu + \delta_\nu^\mu c_\lambda\right), \tag{2.18}$$

with, to begin with, arbitrary vectors b^μ and covectors c_ν. In the local inertial system with $\Pi_{\nu\lambda}^\mu(e) = 0$, the light rays are straight, but not all $K_{\nu\lambda}^\mu(e)$ have to be zero.

Next we perform a 'natural' sharpening of the compatibility between conformal and projective structure which is also implicitly contained in the work of EPS.

Axiom C_2 *A lightlike vector T^μ stays lightlike not only by parallel transport along itself (null geodesic), but also by parallel transport along an arbitrary curve $x^\mu(t)$ with tangent vector $S^\mu = \mathrm{d}x^\mu/\mathrm{d}t$.*
In mathematical terms, Axiom C_2 has the form

$$\frac{\mathrm{d}}{\mathrm{d}t}(g_{\mu\nu} T^\mu T^\nu) = S^\lambda g_{\mu\nu,\lambda} T^\mu T^\nu + 2 g_{\mu\nu} T^\mu \frac{\mathrm{d}T^\nu}{\mathrm{d}t} = 0, \tag{2.19}$$

for $g_{\mu\nu} T^\mu T^\nu = 0$. For this special parallel transport of T^μ, we make the ansatz $\mathrm{d}T^\nu/\mathrm{d}t = -A_{\lambda\kappa}^\nu S^\lambda T^\kappa$, without additional terms proportional to S^ν and T^ν

[compare with (2.15) of Sect. 2.3], and call $A_{\lambda\kappa}^{\nu}$ the *affine connection*. Of course, this special connection is also related to the conformal connection $K_{\lambda\kappa}^{\nu}$ by

$$A_{\lambda\kappa}^{\nu} = K_{\lambda\kappa}^{\nu} + (g_{\lambda\kappa}b^{\nu} + \delta_{\kappa}^{\nu}c_{\lambda} + \delta_{\lambda}^{\nu}c_{\kappa}). \tag{2.20}$$

Inserting (2.20) and the expression for $K_{\lambda\kappa}^{\nu}$ [(2.2) from Sect. 2.2] into (2.19) leads to

$$2T_{\nu}S^{\nu}\left[(b_{\mu} + c_{\mu})T^{\mu}\right] = 0.$$

Since $T_{\nu}S^{\nu}$ is not generally zero, and since there exists a basis of four lightlike vectors T^{μ} at the event e, we get $b_{\mu} = -c_{\mu}$, and (2.20) simplifies to

$$A_{\lambda\kappa}^{\nu} = K_{\lambda\kappa}^{\nu} + (\delta_{\lambda}^{\nu}c_{\kappa} + \delta_{\kappa}^{\nu}c_{\lambda} - g_{\lambda\kappa}c^{\nu}). \tag{2.21}$$

The term $g_{\lambda\kappa}c^{\nu}$ forbids the application of projective transformations to $A_{\lambda\kappa}^{\nu}$, i.e., the above 'natural condition' fixes a unique affine parameter t (up to linear transformations) on any curve $x^{\mu}(t)$. Following EPS, a manifold $\mathcal{M}$ with conformal and projective structures, these being compatible in the sense of Axioms C_1 and C_2, is called a *Weyl structure*.

It is particularly satisfying that there exist simple geometric constructions (and physical Gedanken experiments) for this affine parameter t along the path of a free particle P_1. In all these proposals, one aims to construct a neighboring parallel P_2 to P_1. Light signals reflected between P_1 and P_2 then serve as an affine timekeeper, or as a *geometrodynamical clock*. [In a practical realization, the momentum transfer of the light signal would have the effect that P_1 and P_2 drift apart. However, for 'good realizations', this effect is only noticeable after 10^{13} clock ticks (Harvey 1976).] What is presumed to be the simplest parallel construction has been given in Castagnino (1968), a construction which essentially realizes the Desargues figure in a special case (see Fig. 2.5, and compare with Fig. 1.2 of Sect. 1.4). One constructs the parallel to P_1 through an event B by starting with an event A on P_1 in the past of B (in the interior of the backward light cone of B). A and B then define a free particle path P. From A light is sent to an arbitrary neighboring event A' and reflected back to A'' on P_1. One chooses e on P such that A' and A'' lie in the past of e, and constructs the free particle paths $P' = A'e$ and $P'' = A''e$. One then sends light from B to B' on P' and reflects it back to B'' on P''. If the whole construction is confined to an appropriate neighborhood $\mathcal{U}$ of A and e, the local Desargues theorem (Axiom P_2 of Sect. 2.3) is applicable, i.e., the free particle path P_2 connecting B and B'' is parallel to P_1 (does not meet P_1 in $\mathcal{U}$), because the light ray BB' results from AA' by parallel transport along P, and similarly $B'B''$ from $A'A''$ along P''. Due to the local isotropy of the light propagation and of the projective structure of free particles, the affine parameter t defined by light reflection between P_1 and P_2 is also independent of the specific parallel P_2 one has constructed. (A concrete approximate realization of a geometrodynamical clock is given, e.g., by light or radar reflection between an Earth station and a geostationary satellite.)

Fig. 2.5 The construction of two timelike parallels P_1, P_2 according to the Desargues theorem. Light signals reflected between P_1 and P_2 then define a geometrodynamical clock

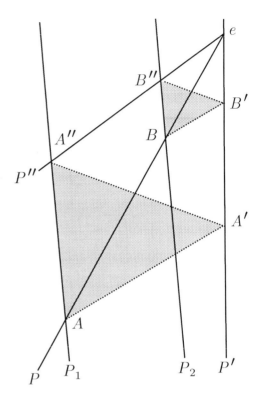

Quite interesting is a quasi-global extension of this geometrodynamical clock. Here the paths P_1 and P_2 do not generally remain parallel (no distant parallelism!), and the light reflected between them defines a time parameter $\tilde{t}$ deviating from the affine parameter t on P_1. In order to prolong the parameter t, one has to construct the parallels to P_1 at events $A_1, A_2, \ldots$ (in the neighborhood of A) anew at each event. However, the resulting parallel $\hat{P}$ is then no longer a global geodesic, and its deviations, and the differences between t and $\tilde{t}$, are governed by the equation of geodesic deviation [compare with (2.8) of Sect. 2.2]:

$$\frac{d^2}{dt^2}\delta x^\mu = Q^\mu_{\nu\rho\sigma}\frac{dx^\nu}{dt}\frac{dx^\rho}{dt}\delta x^\sigma , \qquad (2.22)$$

where $Q^\mu_{\nu\rho\sigma}$ is the curvature tensor constructed from the affine connection $A^\mu_{\nu\lambda}$. In this way the curvature tensor can, in principle, also be measured by geometrodynamical clocks.

The Weyl and affine structures do not yet generally induce a metric structure on $\mathcal{M}$, although we have defined a unique affine parameter on geodesics, and have already spoken of clocks. However, a metric is realized in $\mathcal{M}$ only if for any pair p, q of neighboring events there exists a unique distance function $d(p, q)$ with appropriate properties like the triangle inequality. In contrast, for the affine parameter t on the geodesic P_1, it is not yet guaranteed that by its 'natural transport' with light rays (Einstein's synchronization rule), it transforms to the affine parameter on P_2. In the local coordinate system with $K^\mu_{\nu\lambda}(e) = 0$, the curvature tensor of (2.22) has the structure $Q_{\mu\nu\rho\sigma} = \hat{R}_{\mu\nu\rho\sigma} + g_{\mu\nu} F_{\rho\sigma}$, where the tensor $\hat{R}_{\mu\nu\rho\sigma}$, being antisymmetric in μ, ν, differs from the tensor in (2.8) of Sect. 2.2 only by trace terms with factors $c_\nu c_\rho$ and $c_{\nu,\rho}$, and where $F_{\rho\sigma} = c_{\sigma,\rho} - c_{\rho,\sigma}$. With $dx^\mu/dt = T^\mu$, (2.22) then reads

$$\frac{d^2}{dt^2}\delta x^\mu = \hat{R}^\mu_{\nu\rho\sigma} T^\nu T^\rho \delta x^\sigma + T^\mu (F_{\rho\sigma} T^\rho \delta x^\sigma) \,, \tag{2.23}$$

where the first term on the right-hand side is, due to the antisymmetry of $\hat{R}_{\mu\nu\rho\sigma}$, orthogonal to T^μ. We now state a still sharper, but again 'natural' extension of the compatibility between conformal and projective structure.

Axiom C_3 *If the affine parameter t of a geodesic P_1 is transferred by light reflection to a neighboring parallel geodesic P_2, the result is the affine parameter of P_2.*
According to this axiom, a vector δx^μ between P_1 and P_2 which was orthogonal to T^μ at one place, has to stay orthogonal during the transport along P_1, i.e., in (2.23), the term proportional to T^μ has to vanish, with the consequence $F_{\rho\sigma} = 0$. But then the vector c_σ has the form of a gradient $(\frac{1}{2}\log\Omega^2)_{,\sigma}$, and the relation between $A^\mu_{\nu\lambda}$ and $K^\mu_{\nu\lambda}$ reads

$$A^\mu_{\nu\lambda} = K^\mu_{\nu\lambda} + \frac{1}{2\Omega^2}\left(\delta^\mu_\nu \Omega^2_{,\lambda} + \delta^\mu_\lambda \Omega^2_{,\nu} - g_{\nu\lambda}\Omega^{2,\mu}\right) \,. \tag{2.24}$$

According to Sect. 2.2, the additional Ω^2-dependent terms can be eliminated by a gauge transformation, leading to $A^\mu_{\nu\lambda} = K^\mu_{\nu\lambda}$, and we end up with (up to a constant factor) a unique *metric* and its appertaining connections, the so-called *Christoffel symbols* $\Gamma^\mu_{\nu\lambda}$. In summary, the manifold with Weyl structure becomes a semi-Riemannian manifold.

A popular alternative to Axiom C_3 is the *Riemann axiom*, according to which the transport of the affine parameter or of the geometrodynamical clock is independent of the path taken between a starting point and an end point. [Still another route was recently taken in Matveev and Trautman (2014). These authors reach the metric structure in one step by a new, extended compatibility criterion between conformal and projective structure.] The Riemann axiom is, at least indirectly, open for an experimental test. Although the geometrodynamical clocks according to Marzke–Wheeler, Kundt–Hoffmann, and Castagnino are difficult to realize experimentally, 'gravitational clocks' as given by ephemeris time, solar time due to the Earth's rotation, and pulsar timing are based essentially on free bodies and light rays. Now almost by definition, such clocks are not transportable. However, as stated at the beginning of Sect. 2.2, some of these astrophysical clocks coincide with the best laboratory clocks, in particular atomic clocks, with accuracy up to 10^{-15}. The latter are of course transportable, and their independence of the transport route or of their 'prehistory' has been tested with an accuracy up to 10^{-10} (Hafele and Keating 1972).

Historically, it is remarkable that H. Weyl in Weyl (1918b) postulated a non-vanishing tensor $F_{\rho\sigma}$ (which he called 'Streckenkrümmung') in his attempt to unify gravitation and electromagnetism into one geometric theory. Besides the transport dependence of clocks and rulers, which was immediately criticized by Einstein, such a theory must also be rejected because electromagnetism is not a universal, geometric force: there is no substitute for the equivalence principle, and charged particles with different mass-to-charge-ratio move on different paths. (Compare with the statement in Sect. 1.4 that the Lorentz force of electrodynamics is the simplest Minkowski force that cannot be transformed away.) However, as already mentioned in Sect. 2.1, Weyl's ideas, and in particular his gauge principle, have witnessed a glorious revival in modern elementary particle theory.

Another experimental test which is intimately connected with the compatibility between conformal and projective structure is the *gravitational redshift*. Here typically, e.g., in the Pound–Rebka experiment (Pound and Rebka 1960), the characteristic frequencies ν_0 and ν_h of atomic nuclei A at the Earth's surface and B at height h are compared. In such relatively local experiments, special relativity is sure to be a sufficient approximation, and spacetime curvature is irrelevant. The question is only whether the local inertial system in which the light velocity has the universal value c is firmly connected to (a) the Earth, or to (b) a freely falling particle P, with compatibility only in the latter case. Since particles with constant acceleration g describe hyperbolas in a Minkowski diagram, we have the two pictures of Fig. 2.6. Obviously, only the alternative (b) gives the correct redshift $\Delta\nu/\nu \approx gh/c^2$.

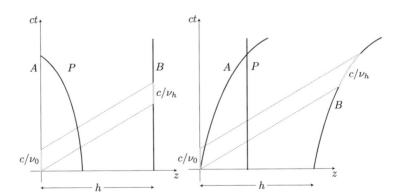

Fig. 2.6 Two Minkowski diagrams for the free fall of particle P to Earth. *Left*: Free fall and the emission of consecutive photons in the rest system of the Earth [situation (a)]. *Right*: Likewise, in the rest system of P [situation (b)]

Returning to the mathematics of a metric manifold $\mathcal{M}$, we have, as already mentioned, the formula for the Christoffel symbols:

$$\Gamma^{\mu}_{\nu\lambda} = A^{\mu}_{\nu\lambda} = K^{\mu}_{\nu\lambda} = \frac{1}{2}g^{\mu\rho}(g_{\rho\nu,\lambda} + g_{\rho\lambda,\nu} - g_{\nu\lambda,\rho}) \,. \tag{2.25}$$

This implies immediately

$$g_{\mu\nu;\lambda} = 0 \,, \tag{2.26}$$

which is presumably the most compact mathematical expression for the compatibility between conformal and projective structure. For the Riemann curvature tensor $R^{\mu}_{\nu\rho\sigma}$, we have in the local inertial system,

$$R^{\mu}_{\nu\rho\sigma;\kappa} = R^{\mu}_{\nu\rho\sigma,\kappa} = \Gamma^{\mu}_{\nu\sigma,\rho\kappa} - \Gamma^{\mu}_{\nu\rho,\sigma\kappa} \,, \tag{2.27}$$

and this immediately implies the *Bianchi identities* (Bianchi 1902), viz.,

$$R^{\mu}_{\nu\rho\sigma;\kappa} + R^{\mu}_{\nu\sigma\kappa;\rho} + R^{\mu}_{\nu\kappa\rho;\sigma} = 0 \,, \tag{2.28}$$

and by taking the trace, the *contracted Bianchi identities*, viz.,

$$R^{\mu}_{\nu\rho\sigma;\mu} = R_{\nu\sigma;\rho} - R_{\nu\rho;\sigma} \,. \tag{2.29}$$

Due to the symmetries of the Riemann tensor, both (2.28) and (2.29) comprise 20 independent equations, and the contracted Bianchi identities are even equivalent to the more complicated looking original Bianchi identities. Equation (2.29) can also be transcribed into a (differential!) relation between the Weyl tensor $C^{\mu}_{\nu\rho\sigma}$ and the Ricci tensor $R_{\nu\sigma} = R^{\mu}_{\nu\mu\sigma}$ and the scalar curvature $R = g^{\nu\sigma}R_{\nu\sigma}$:

$$C^{\mu}_{\nu\rho\sigma;\mu} = \frac{1}{2}(R_{\nu\sigma;\rho} - R_{\nu\rho;\sigma}) - \frac{1}{12}(g_{\nu\sigma}R_{,\rho} - g_{\nu\rho}R_{,\sigma}) \,. \tag{2.30}$$

Historically, it is remarkable that these nice and important Bianchi identities were unknown to Einstein, Hilbert, and Weyl, and to the rest of the physics community investigating general relativity. And this lack of knowledge was a decisive obstacle for an earlier and more elegant derivation of the correct Einstein field equations, i.e., of the way in which matter produces gravitational fields in the form of a spacetime curvature (see Sect. 3.1). For some interesting history concerning the role of the Bianchi identities in general relativity, and in particular in Pauli's famous original encyclopedia article of 1921 and in its English reedition of 1958, see the Pauli biography by C. Enz (2002, pp. 30–35).

References

Abraham, R., Marsden, J.E., Ratiu, T.: Manifolds, Tensor Analysis, and Applications. Springer, New York (1988)

Bianchi, L.: Sui simboli a quattro indici e sulla curvatura di Riemann, Rend. della R. Acc. dei Lincei **11**, 3–7 (1902)

Castagnino, M.: Some remarks on the Marzke–Wheeler method of measurement. Nuovo Cimento B **54**, 149–150 (1968)

Castagnino, M.A.: The Riemannian structure of space-time as a consequence of a measurement method. J. Math. Phys. **12**, 2203–2211 (1971)

Coleman, R.A., Korte, H.: Jet bundles and path structures. J. Math. Phys. **21**, 1340–1351 (1980)

Ehlers, J.: The nature and structure of spacetime. In: Mehra, J. (ed.) The Physicist's Conception of Nature, pp. 71–91. Reidel, Dordrecht (1973a)

Ehlers, J.: Survey of general relativity theory. In: Israel, W. (ed.) Relativity, Astrophysics and Cosmology, pp. 1–125. Reidel, Dordrecht (1973b). Esp. Sect. 1.2. Newtonian space-time, Mechanics, and Gravity Theory

Ehlers, J.: Hermann Weyl's contributions to the general theory of relativity. In: Deppert, W., et al. (eds.) Exact Sciences and their Philosophical Foundations, pp. 83–105. Peter Lang, Frankfurt (1988)

Ehlers, J.: Einführung in die Raum-Zeit-Struktur mittels Lichtstrahlen und Teilchen. In: Audretsch, J., et al. (eds.) Philosophie und Physik der Raum-Zeit, pp. 145–162. BI-Wissenschaftverlag, Mannheim (1988)

Ehlers, J., Köhler, E.: Path structures on manifolds. J. Math. Phys. **18**, 2014–2018 (1977)

Ehlers, J., Schild, A.: Geometry in a manifold with projective structure. Commun. Math. Phys. **32**, 119–146 (1973)

Ehlers, J., Pirani, F.A.E., Schild, A.: The geometry of free fall and light propagation. In: O'Raifertaigh, L. (ed.) General Relativity. Papers in Honour of J.L. Synge, pp. 63–84. Clarendon Press, Oxford (1972). Republished as 'Golden Oldie' in Gen. Relativ. Gravit. **44**, 1587–1609 (2012), together with an editorial note by A. Trautman in Gen. Relativ. Gravit. **44**, 1581–1586 (2012)

Enz, C.P.: No Time to Be Brief. A Scientific Biography of Wolfgang Pauli. Oxford University Press, Oxford (2002)

Fermi, E.: Sopra i fenomeni che avvengono in vicinanza di una linea oraria. Atti Reale Accad. Naz. dei Lincei, Rendiconti Cl. sci. fis., mat. e nat. **31**, 51–53 (1922)

Frauendiener, J., Friedrich, H. (eds.): The Conformal Structure of Space-Times. Lecture Notes in Physics, vol. 604. Springer, Berlin (2002)

Hafele, J.C., Keating, R.E.: Around-the-world atomic clocks: observed relativistic time gains. Science **177**, 168–170 (1972)

Harvey, A.: Photon clocks. Gen. Relativ. Gravit. **7**, 891–893 (1976)

Hawking, S., Ellis, G.F.R.: The Large Scale Structure of Space-Time. Cambridge University Press, Cambridge (1973)

Heilig, U., Pfister, H.: Characterization of free fall paths by a global or local Desargues property. J. Geom. Phys. **7**, 419–446 (1990)

Helmholtz, H.: Über die Tatsachen, die der Geometrie zu Grunde liegen. Nachr. Ges. Wiss. Göttingen, 193–221 (1868)

Heusler, M.: Black Hole Uniqueness Theorems. Cambridge University Press, Cambridge (1996)

Hobbs, G., et al.: Development of a pulsar-based time-scale. Mon. Not. R. Astron. Soc. **427**, 2780–2787 (2012)

Jost, J.: Bernhard Riemann. Über die Hypothesen, welche der Geometrie zu Grunde liegen. Springer, Berlin (2013)

Kerr, R.P.: Gravitational field of a spinning mass as an example of algebraically special metrics. Phys. Rev. Lett. **11**, 237–238 (1963)

Laugwitz, D.: Differential and Riemannian Geometry. Academic, New York (1965)

Levi-Civita, T.: Nozione di parallelismo in una varietá qualunque e consequente specificazione geometrica della curvatura Riemanniana. Rend. Circ. Mat. Palermo **42**, 173–205 (1917)

Longo, M.J.: Tests of relativity from SN1987A. Phys. Rev. **36**, 3276–3277 (1987)

Matveev, V.S., Trautman, A.: A criterion for compatibility of conformal and projective structures. Commun. Math. Phys. **329**, 821–825 (2014)

Meinel, R.: A physical derivation of the Kerr–Newman black hole solution (2013). E-print. arXiv: 1310.0640 [gr-qc]

Misner, C.W., Thorne, K.S., Wheeler, J.A.: Gravitation. Freeman, San Francisco (1973)

Penrose, R.: Conformal treatment of infinity. In: deWitt, C.M., deWitt, B. (eds.) Relativity, Groups and Topology. Gordon and Breach, New York (1964)

Penrose, R.: Gravitational collapse and space-time singularities. Phys. Rev. Lett. **14**, 57–59 (1965)

Penrose, R.: Singularities and time-asymmetry. In: Hawking, S.W., Israel, W. (eds.) General Relativity: An Einstein Centenary. Cambridge University Press, Cambridge (1979)

Pirani, F.A.E.: A note on bouncing photons. Bull. L'Academie Polonaise des Sciences, Ser. Sci. Math. Astr. Et Phys. **13.3**, 239–242 (1965)

Pirani, F.A.E.: Building space-time from light rays and free particles. In: Symposia Mathematica, vol. 12, pp. 67–83. Academic, London (1973)

Pirani, F.A.E., Schild, A.: Conformal geometry and the interpretation of the Weyl tensor. In: Hoffmann, B. (ed.) Perspectives in Geometry and Relativity, pp. 291–309. Indiana University Press, Bloomington (1966)

Pound, R.V., Rebka, G.A.: Apparent weight of photons. Phys. Rev. Lett. **4**, 337–341 (1960)

Sachs, R.: Gravitational waves in general relativity VI. The outgoing radiation condition. Proc. R. Soc. Lond. **A264**, 309–338 (1961)

Schlamminger, S., et al.: Test of the equivalence principle using a rotating torsion balance. Phys. Rev. Lett. **100**, 041101 (2008)

Stephani, H., Kramer, D., MacCallum, M., Hoenselaers, C., Herlt, E.: Exact Solutions of Einstein's Field Equations. Cambridge University Press, Cambridge (2003)

Synge, J.L.: Relativity: The General Theory. North Holland, Amsterdam (1966)

Szekeres, P.: The gravitational compass. J. Math. Phys. **6**, 1387–1391 (1965)

Weyl, H.: Reine Infinitesimalgeometrie. Math. Zs. **2**, 384–411 (1918a)

Weyl, H.: Gravitation und Elektrizität, pp. 465–480. Sitzb. d. Preuss. Akad. d. Wiss., Berlin (1918b)

Weyl, H.: Raum, Zeit, Materie, 1st edn. Springer, Berlin (1919); 5th edn. Springer, Berlin (1922); 8th edn. Springer, Berlin (1993)

Weyl, H.: Zur Infinitesimalgeometrie: Einordnung der projektiven und der konformen Auffassung. Nachr. Königl. Ges. Wiss. Göttingen, Math.-Phys. Kl, 99–112 (1921)

Weyl, H.: Die Einzigartigkeit der Pythagoreischen Maßbestimmung. Math. Zs. **12**, 114–146 (1922)

Weyl, H.: Geometrie und Physik. Die Naturwissenschaften **19**, 49–58 (1931)

Zeeman, E.C.: The topology of Minkowski space. Topology **6**, 161–170 (1967)

Chapter 3
Einstein's Field Equations, Their Special Mathematical Structure, and Some of Their Remarkable Physical Predictions

3.1 Many Different Routes to Einstein's Field Equations

In this section we first discuss some essential steps which Einstein took on his long and tedious route to the final formulation of general relativity, a route which is characterized by revolutionary and glorious ideas, but also (with hindsight) by distressing physical and mathematical errors and misjudgements. Then we report on the many different routes to general relativity which have been investigated up to today, and we argue that this quasi-uniqueness and 'inevitability' of general relativity is a special strength of this relativistic theory of gravitation. (Nevertheless, we cannot speak of a logical 'derivation' of the theory, since this would be impossible for any theory applying to new types of physical phenomena.)

The first very important step in the direction of a new and relativistic theory of gravitation is contained in Einstein's overview of special relativity in his article for the *Jahrbuch der Radioaktivität und Elektronik* (Einstein 1907). Here Einstein asks (on p. 454, in the last part of this article) whether the relativity principle can be extended to systems which are accelerated relative to each other. In particular, he postulates what became known later as the *Einstein equivalence principle*, according to which a reference system in uniform translational acceleration and a rest system with a homogeneous gravitational field are completely equivalent concerning all nongravitational physical phenomena. Later, in a draft for an invited article for the journal *Nature* in 1921, Einstein called this equivalence principle "the happiest thought of my life" [see the Einstein biography by Pais (1982, pp. 177–178)]. Already in 1884 H. Hertz in a lecture on the constitution of matter, stressed the equality of inertial and gravitational mass as a remarkable fact of nature (Hertz 1884, pp. 121–122):

> Most miraculous is the connection between the gravitation of matter and its inertia. We see that any of two quantities of matter with the same inertia will exert the same gravitational effect on each other, irrespective of the substances they are made of. [...] And in reality we do have two properties before us, two most fundamental properties of matter, which must

© Springer International Publishing Switzerland 2015
H. Pfister, M. King, *Inertia and Gravitation*, Lecture Notes in Physics 897,
DOI 10.1007/978-3-319-15036-9_3

be thought of as being completely independent of each other, but which, in our experience, and only in our experience, appear to be exactly equal. This correspondence must mean much more than a mere miracle; it demands an explanation. We may suppose that a simple and intelligible explanation is also possible, and that such an explanation will provide an extensive insight into the constitution of matter.

On the basis of his equivalence principle, Einstein argues in Einstein (1907) that gravitational fields will influence the frequency of clocks and of spectral lines, that light rays are curved, and that the law $E = mc^2$ applies also to the gravitational mass. From today's perspective, it has to be said that the equivalence principle still plays an important but also very delicate role [see, e.g., the articles Ohanian (1977) and Norton (1985)]. First, it has to be stressed that a precise equivalence principle can at best be valid locally, as was also stressed by Einstein in Einstein (1912b). A global homogeneous gravitational field is unrealistic in any case, and a Minkowskian reference system in arbitrary acceleration remains flat, whereas genuine gravitational fields are represented by curved manifolds (but see our quasiglobal equivalence principle in Sects. 3.3 and 4.2). But even a local equivalence principle has its problems, because, besides other difficulties, in the final theory of general relativity, test particles with spin move on paths deviating (slightly) from geodesics, as was first proven in Papapetrou (1951). An extremely negative point of view concerning the equivalence principle is taken by J. Synge in the preface to his textbook on general relativity (Synge 1960, pp. IX–X):

> The Principle of Equivalence performed the essential office of midwife at the birth of general relativity, but, as Einstein remarked, the infant would never have got beyond its long-clothes had it not been for Minkowski's concept. I suggest that the midwife be now buried with appropriate honours and the facts of absolute space-time be faced.

This drastic formulation surely contains a grain of truth, and in the final mathematical formulation of general relativity the equivalence principle is no longer really necessary. However, it is convenient, and it has been general practice in the physics literature up to today, to retain various kinds of equivalence principles as essential pillars of general relativity, which allow one to understand, or even predict, some mathematical consequences of this theory heuristically by qualitative physical reasoning.

The principle makes it plausible that the theory of gravity has to be a metric theory, that gravity couples universally and minimally to all physical systems, and that locally all physical laws attain their special relativistic form and are independent of the position and velocity of the relevant measuring equipment. In today's textbook literature [see in particular Will (1993)], there usually appear, besides the Einstein equivalence principle, a so-called *weak equivalence principle*, expressing essentially the equality of inertial and gravitational mass, and a *strong equivalence principle*, extending the Einstein equivalence principle to systems with nontrivial and strong gravitational fields, with important consequences for, e.g., the Earth–Moon–Sun system (Nordtvedt effect), and for strongly bound astrophysical systems. In summary, the equivalence principle makes it plausible that a relativity theory generalized to accelerated systems is automatically a theory of gravity, and that conversely, a satisfying theory of gravity can only be formulated through a generalized relativity postulate. The textbook (Misner et al. 1973, p. 207) compares the role of the equivalence principle in general relativity with the role of the correspondence principle in quantum mechanics:

> The vehicle that carries one from classical mechanics to quantum mechanics is the correspondence principle. Similarly, the vehicle between flat spacetime and curved spacetime is the equivalence principle.

A second important step on Einstein's route to general relativity is of course his introduction of a non-Euclidean geometry, and in particular, a four-dimensional Riemannian geometry $g_{\mu\nu}(x^\lambda)$ of spacetime. The beginning of such thoughts can be traced back to a letter to A. Sommerfeld on 29 September 1909 (Klein et al. 1993, Doc. 179), in which he writes:

> The treatment of the uniformly rotating rigid body seems to me to be of great importance on account of an extension of the relativity principle to uniformly rotating systems along analogous lines of thought to those I tried to carry out for uniformly accelerated translation in the last section of my paper published in the Zeitschrift für Radioaktivität (Einstein 1907).

At this time the problem of rotating rigid bodies in special relativity was being actively pursued by quite a number of theoretical physicists (M. Born, P. Ehrenfest, G. Herglotz, T. Kaluza, F. Noether, and others), initiated mainly by a model for a rigidly rotating electron in Born (1909). Ehrenfest (1909) hinted at a contradiction (later called the Ehrenfest paradox) that would occur for all rigid bodies rotating around a fixed axis: whereas the periphery of such a body suffers a length contraction, the radius stays invariant. This statement was immediately interpreted by the above authors to the effect that a rotating disk has to be described by a non-Euclidean geometry. Einstein discussed these problems with Born, Sommerfeld, and others in 1909 (e.g., at the Naturforscher-Versammlung in Salzburg), but first published about them in Einstein (1912a). And in 1916, in his first review article on the final version of general relativity, Einstein says: "Hence Euclidean geometry does not apply to [the system of the rotating disk]". In spring 1913, Einstein took a big and quite surprising step by introducing for fundamental spacetime (and not only for rigid bodies) a four-dimensional Riemannian geometry $g_{\mu\nu}(x^\lambda)$ in the so-called Entwurf paper (Einstein and Grossmann 1913), although up to this time he had, even in special-relativistic calculations, not used the elegant four-dimensional formalism of Minkowski.

In retrospect it has, however, to be stressed that the above 'derivation' of a non-Euclidean geometry on a rotating rigid disk is not tenable: length contraction is a phenomenon relating two inertial systems, and by no means an invariant statement. At least since 1959 it has been known that whether and what length contraction appears in the observation of a rod depends decisively on the details of the experiment. Moreover, we have seen in Sects. 2.2 and 2.3 that a two-dimensional manifold (e.g., of the rotating disk) has neither a conformal nor a projective curvature. From today's perspective, presumably the most convincing argument that a relativistic gravity theory fulfilling the Einstein equivalence principle necessitates a curved Riemannian geometry comes from one of the alternative routes to the

Einstein field equations to be discussed below as route 5: beginning with a special-relativistic linear field theory for a massless spin-2 field, the equivalence principle (the energy of this field couples back, generating a nonlinear correction to the original field equation) starts an iterative process, resulting uniquely in Einstein's field equations in a curved manifold, and leaving the original Minkowski metric as an idle artefact.

But even before the Entwurf paper, in his year in Prague, Einstein derived important physical results concerning a future relativistic gravity theory satisfying his equivalence principle. In Einstein (1911), he discussed in more detail than in Einstein (1907) the gravitational effects of energy and the connection between the redshift and the gravitational potential. He also discussed the variability of the light velocity in a gravitational field, and he announced with particular pleasure that the deflection of light passing by the rim of the Sun would be approximately 1 arcsec, and might therefore be measurable. In Einstein (1912a), he developed the first, still linear and scalar, relativistic gravity theory for static gravitational fields which was compatible with the equivalence principle for uniform translational acceleration. In Einstein (1912b), he extended this to a nonlinear theory (due to the energy of the gravitational field), discussed in detail the interactions between electromagnetic fields and this static gravity field, as well as thermodynamic aspects, and finally derived a Lagrangian form for the equations of motion in this gravity field. In Einstein (1912c), he applied this theory to a special model: he introduced a spherical mass shell (still useful today in the final version of general relativity, as described in Sect. 4.2), and argued that a linear acceleration of this shell would induce a (small) acceleration of test bodies inside the shell. This was the first calculation of a (Machian) dragging effect in a gravity theory. (We will come back to this paper as a stimulus for more detailed dragging calculations in general relativity in Sect. 4.2.)

In the period August 1912 to May 1913, there are no publications by Einstein on gravity theory. But there is the quite important Zurich notebook (Klein 1995, pp. 192–269) covering presumably just this period. The evaluation of this notebook by historians of science (see, e.g., Norton 1984) has resulted in decisive clarifications concerning Einstein's route to the Entwurf paper (Einstein and Grossmann 1913), and even to the final theory of general relativity. For instance, the notebook together with his earlier work in Einstein (1907, 1911, 1912a,b,c) reveals that Einstein gradually propounded the following heuristic principles for a future relativistic gravity theory, although to the best of our knowledge he nowhere formulated things in this systematic and explicit form:

- It should be based on a Riemannian geometry $g_{\mu\nu}(x^\lambda)$ (appearing already on the first pages of the notebook), and this should be the only gravitational field variable.
- The field equations for $g_{\mu\nu}$ should be of second order and nonlinear.
- The equivalence principle should be realized as completely as possible, and therefore the field equations should be covariant with respect to as many coordinate transformations as possible, including the transformation to accelerated, and if possible to rotating reference systems.

- The source of the gravitational field must be the whole (symmetric) energy–momentum tensor $T_{\mu\nu}$ of all matter and electromagnetic fields, i.e., we are dealing with a tensorial gravity theory. (According to special relativity, the energy density $\varrho = T_{00}$ is necessarily connected to the momentum density T_{0i}, and according to the momentum conservation law, the momentum density is connected to a momentum–current density T_{ik}, including also the pressure components T_{ii}. At that time, this whole complex of the energy–momentum tensor $T_{\mu\nu}$ was also already well known from Maxwell's relativistic electrodynamics.) Scalar gravity theories, as investigated around the same time by G. Nordström, M. Abraham, and G. Mie, have to be rejected because their scalar source T_μ^μ would be zero for electromagnetic radiation, whereas this clearly has energy.
- The field equations have to be compatible with energy and momentum conservation for the whole system composed of matter and electromagnetic and gravitational fields.
- In the appropriate limits, the new theory has to coincide with Newtonian gravity, and with the static relativistic theory of Einstein (1912b).

The notebook also gives us an insight into the way Einstein gradually came to understand Riemannian geometry with the help of his friend the mathematician M. Grossmann. It was a very difficult and unusual field for a physicist to study at this time. In this connection the notebook reveals that Einstein and Grossmann even considered setting the Riemann tensor and the Ricci tensor for the gravitational field proportional to the energy–momentum tensor. In this way, by the spring of 1913, they had already come "within a hair's breadth" (Norton 1984) of the correct field equations for the final version of general relativity of November 1915, at least in the linearized form. (This judgement is literally confirmed by Einstein in his first paper Einstein (1915a) of November 1915; see below.) However, they rejected this ansatz because they (erroneously) believed that it did not fulfil the last principle of the above list. [They did not recognize that the Newtonian limit does not have the standard form $\Delta\Phi = 4\pi\rho$ in all reference systems. Furthermore, they did not realize that static relativistic gravitational fields do not always have to reduce to one potential, and are not always spatially flat, as in Einstein (1912b).]

The end of the notebook then indicates alternative field equations which are more explicitly motivated and formulated in the Entwurf paper (Einstein and Grossmann 1913). (Since these equations are quite complicated and no longer play any role in today's gravity theory, we shall not give their detailed form.) However, Einstein stuck to these equations for more than 2 years, and tried to justify and 'derive' them, one can say, with growing despair. For instance, he found that these equations are not generally covariant, not even for transformations to rotating reference systems. Many of his following arguments are based on the hypothesis "that the field be mathematically completely determined by matter", e.g., in a letter to L. Hopf of 2 November 1913, published in Klein et al. (1993, Doc. 480). ('Completely' or 'uniquely' should here be understood in the physical sense, i.e., it would have been clear to Einstein that the $g_{\mu\nu}$ change in going to other coordinates.)

Nevertheless, we know today that the above hypothesis is fundamentally wrong: mathematically, the manifold of different solutions of given partial differential equations of second order goes far beyond the choice of coordinates. Physically, we know that, e.g., for the vacuum ($T_{\mu\nu} \equiv 0$), besides the Minkowski solution, there exist many different solutions of the Einstein equations: Schwarzschild, Kerr, exact wave solutions, etc. In contrast, from the above hypothesis, Einstein derived that generally covariant field equations are physically unacceptable, e.g., by the 'notorious' hole argument [change of $g_{\mu\nu}$ only in a hole region where $T_{\mu\nu}$ is zero; for more details see Norton (1984)]. However, besides such erroneous conclusions from the Entwurf theory, Einstein also reached (at least qualitatively) interesting, new, and lasting consequences of such a nonlinear and tensorial relativistic gravity theory. For instance, in Einstein (1914) he says: "We have no means of distinguishing a 'centrifugal field' from a gravitational field. [...] we may consider the centrifugal field to be a gravitational field." Indeed, as we will show explicitly in Sect. 4.2 and in Appendix B, there are models of slowly rotating mass shells in general relativity which produce a correct centrifugal field as a gravitational field.

This situation lasted until autumn 1915, when Einstein came to realize the decisive errors and misconceptions in the Entwurf paper, and in the later publications based on it. In an admirably short time in three papers of November 1915, he arrived at the correct (Einstein) field equations of the final version of general relativity. In Einstein (1915a) of 4 November, he returned to (almost) general covariance, although with the restriction that the trace of the metric tensor should remain unchanged, with the consequence that the theory applies only to matter with $T_\mu^\mu = 0$. His field equations set the Ricci tensor proportional to the energy–momentum tensor, and he confirms that (in contrast to his statement in 1912–1913) these equations have the correct Newtonian limit. He also explicitly formulates as follows: "In fact we [Einstein and Grossmann] had already at that time [1912–1913] come quite near to the solution of the problem that is given in what follows." In the paper Einstein (1915b) of 18 November, he did an approximate calculation (small deviations up to second order from the Minkowski metric) of the perihelion advance of Mercury, with the agreeable result of 43 arcsec per century. For the deflection of light passing by the Sun he found (thanks to the spacetime curvature) twice the value of his earlier calculation in Einstein (1911). (The fact that he still had the wrong relation between the Ricci tensor and the energy–momentum tensor did not matter because both effects are governed by the vacuum equations.) Finally, in Einstein (1915c) of 25 November, he removed the restriction on the trace of the metric tensor and arrived at a fully covariant theory. Furthermore, he changed the relation between the Ricci tensor and the energy–momentum tensor in such a way that it is compatible with the energy–momentum conservation laws. In this way, he arrived at the Einstein field equations (in today's notation)

$$G_{\mu\nu} = R_{\mu\nu} - \frac{1}{2}Rg_{\mu\nu} = 8\pi T_{\mu\nu} \, . \tag{3.1}$$

(Since he did not know of the Bianchi identities at that time, as already mentioned at the end of Sect. 2.4, he did not see that conservation of energy and momentum followed elegantly and automatically from $G^{\mu\nu}{}_{;\nu} = 8\pi T^{\mu\nu}{}_{;\nu} = 0$.) In a letter to A. Sommerfeld on 9 December 1915 (Schulman 1998, Doc.161), Einstein called (3.1) "the most valuable discovery I have made in my life".

Before finishing our analysis of Einstein's route to general relativity, we would like to make two more remarks. The first concerns D. Hilbert. As is well known, Hilbert published in Hilbert (1916) the correct field equations of general relativity (by a different method, namely an action principle) around the same time as Einstein. In this connection, it has to be said that Hilbert learnt about the status of Einstein's project of a relativistic gravity theory and its mathematical and physical difficulties from six 2-h lectures which Einstein gave in Göttingen in June 1915, and from letter exchanges after this time. Furthermore, a more recent evaluation (Corry et al. 1997) of the publication process of Hilbert's paper (Hilbert 1916) has revealed that the paper submitted on 20 November 1915, 5 days before the submission of Einstein's paper (Einstein 1915c), did not contain the correct field equations. These only appeared in the heavily revised version, published on 31 March 1916. This does not of course imply that Hilbert plagiarized the correct form from Einstein, but it makes it clear that the priority in publishing the correct final version of general relativity lies with Einstein.

The second remark concerns once more the Entwurf paper (Einstein and Grossmann 1913). Although this paper contains no applications of the new theory, such applications were calculated by Einstein with his friend M. Besso in the so-called Einstein–Besso manuscript (Klein 1995, pp. 344–473), written mostly in June 1913. The main part of this manuscript addresses the calculation of the perihelion advance of Mercury, which Einstein had already mentioned as a central goal for a future relativistic gravity theory in a letter to C. Habicht on 24 December 1907 (Klein et al. 1993, p. 82). However, the result in the Entwurf theory was only 18 arcsec per century, as opposed to the observed value of 45 arcsec. Furthermore, Einstein and Besso calculated three so-called dragging effects in this manuscript: the linear dragging of a test mass inside a linearly accelerated mass shell [compare Einstein's paper (Einstein 1912c)], the rotational (Coriolis) dragging inside a rotating mass shell (half the value in the final version of general relativity), and a motion of the planetary nodes in the field of the rotating sun (one fourth of the value in general relativity). It is interesting to note which parts of this manuscript Einstein presented in his brilliant speech at the Naturforscher-Versammlung in Vienna on 21 September 1913 (Einstein 1913), and which parts he omitted. It is also remarkable that all these (non-Newtonian!) gravitational effects already come out qualitatively correctly in the Entwurf theory, although by factors 2–4 smaller than in general relativity. (We shall come back to the above dragging effects, their calculation, and their physical role in general relativity in Sect. 4.2.)

It is a remarkable feature of general relativity that its basic equations, Einstein's field equations, can be motivated or even 'derived' along many alternative routes, based on very different suppositions. In its Box 17.2, the textbook (Misner et al. 1973) (MTW) gives six such different routes. In the following we refer to these routes, to some only very briefly, to others more extensively. In addition, we report on some progress and some new views on this topic, published since 1973.

Route 1 in the list of MTW is closest to Einstein's own approach to his field equations. It asks, and analyzes mathematically, what singles out the Einstein tensor $G_{\mu\nu}$ of (3.1) from other metric tensors. As early as 1916–1922, the following result was indicated in Einstein (1916), and derived in Vermeil (1917), Weyl (1922), and Cartan (1922): a covariant, symmetric, and divergence-free 2-tensor (in order to produce the automatic conservation of energy and momentum), whose coefficients are functions only of the metric and its first and second derivatives, and which is linear in the second derivatives, is necessarily a sum $aG_{\mu\nu} + bg_{\mu\nu}$, with constants a and b (cosmological constant, usually denoted by Λ). Considerably later, the above conditions could be much weakened in Lovelock (1969) and Lovelock (1972): in the physical dimension 4, the symmetry of the tensor, and its linearity in the second derivatives are already a consequence of the other conditions. Still later, it was argued (Aldersley 1977; Navarro and Sancho 2008) with the help of a so-called dimensional analysis that this 'derivation' of the Einstein tensor can be extended to arbitrary dimensions, and to a dependence also on higher derivatives of the metric.

Route 2 in MTW's list is based on a variational principle, and it is in some sense the simplest and quickest route. As remarked above, it was first taken in Hilbert (1916). If the (scalar!) Lagrangian density of this variational principle is linear in the second derivatives of the metric, contains no higher derivatives, and vanishes in flat spacetime, it has, up to a constant factor, to be the scalar curvature invariant R, and variation of the integral $\int R \sqrt{-g} \, d^4x$ with respect to the metric leads uniquely to the Einstein vacuum equations. In detail, this process is somewhat tricky. Since the curvature scalar R contains second derivatives of the metric, one might expect the resulting field equations to be of third order. (And there does not exist a scalar containing only the first derivatives of the metric!) However, it turns out that the 'dangerous terms' in the integral can be combined into a four-divergence which does not contribute to the variation of the integral. Addition of an appropriate matter Lagrangian then results in the full Einstein equations.

According to Palatini (1919) there exists an alternative variation of the above integral, motivated by the Hamiltonian formulation of classical mechanics, such that the metric $g_{\mu\nu}$ and the Christoffel symbols $\Gamma^{\mu}_{\nu\lambda}$ are varied as independent variables. Then, besides Einstein's field equations, the relation between $\Gamma^{\mu}_{\nu\lambda}$ and the first derivatives $g_{\mu\nu,\lambda}$ follows [see (2.25) in Sect. 2.4]. Route 2 is also very appropriate for constructing relativistic gravity theories different from general relativity. For instance, in recent times, the so-called $f(R)$ theories of gravity have seen much activity. Here the scalar curvature invariant R in the variational integral is replaced by some nonlinear function $f(R)$. Such extensions of general relativity are motivated by difficulties with Einstein's theory, mainly in cosmology (dark matter, dark energy, accelerated expansion), and in the problem of quantizing the gravity theory. However, it should be said that until now no preferred function $f(R)$ has shown up, and many models have difficulty reproducing all the observational facts, such as the existence of compact stars and black holes. For reviews of such $f(R)$ theories see Sotirio and Faraoni (2010) and De Felice and Tsujikawa (2010).

Route 3 of MTW is based on data solely on a spacelike slice, the hypersurface of simultaneity $\mathscr{S}$ of an observer with four-velocity **u**. He/she sets the sum of the intrinsic curvature $^{(3)}R$ of $\mathscr{S}$ and the scalar of the extrinsic curvature tensor **K** proportional to the local energy density $\varrho = T_{\mu\nu}u^{\mu}u^{\nu}$, in the form

$$^{(3)}R + (\mathrm{Tr}\mathbf{K})^2 - \mathrm{Tr}\mathbf{K}^2 = 16\pi\varrho .$$

It can then be shown that the validity of this equation on every slice of every spacelike hypersurface $\mathscr{S}$ results in the full four-dimensional Einstein field equations.

Route 4 of MTW proceeds by analogy with the Hamilton–Jacobi formulation of classical mechanics, and is based on the concept of 'superspace' (Wheeler 1962). This is an infinite-dimensional space of (diffeomorphic equivalence classes of) three-geometries $\mathscr{S}(g_{ij})$. The Hamilton–Jacobi equation

$$-\frac{(16\pi)^2}{2g}\left(g_{im}g_{jn} + g_{in}g_{jm} - g_{ij}g_{mn}\right)\frac{\delta\mathscr{S}}{\delta g_{ij}}\frac{\delta\mathscr{S}}{\delta g_{mn}} + {}^{(3)}R = 16\pi\varrho \qquad (3.2)$$

characterizes the three-geometries which fit into a dynamic four-geometry, or which guarantee a constructive interference of the wavefronts of $\mathscr{S}$. It can then be shown (Gerlach 1977) that all ten Einstein field equations follow from (3.2). The concept of superspace has also been the basis for attempts to find a quantum theory of gravity.

In some ways the most surprising and most interesting route to the Einstein field equations is route 5 in the list of MTW. This route was initiated in Gupta (1954) and Kraichnan (1955), was partly worked out in Thirring (1961) and in the 1963 Caltech lectures on gravitation by Feynman (1995), and brought to an elegant conclusion in Deser (1970). The route has its source in (special relativistic) field theory and elementary particle physics. If gravity has any chance of being described in this scheme, one has to start with 'particles' (so-called gravitons) of mass zero and spin two, due to the infinite range and the tensor character of a relativistic gravity theory. Such a relativistic wave equation was first derived in Fierz and Pauli (1939). However, as a linear equation, it violates the equivalence between energy and mass. In order to remedy this inconsistency, the energy of the spin-2-field, being quadratic in the field amplitude, has to couple back as a source of the field equation. In this way, an iterative process (a type of renormalization) is started which has to be continued infinitely, in order to guarantee conservation of energy and momentum. (In the Deser formulation, this infinite series is cut short to one 'simple' self-interaction.) The final result of this process is the full, nonlinear Einstein field equations in a curved manifold, and the fulfillment of the equivalence principle. The original flat space-time metric remains only as an unobservable artefact. As already remarked above, this represents presumably the most convincing argument for a curved Riemannian geometry for any consistent relativistic gravity theory. In the words of Wheeler, "curvature without curvature".

In connection with the different routes to the Einstein field equations, and in particular concerning route 5, a comment on string theory seems to be appropriate.

Sometimes one can read statements like "these theories have the remarkable property of predicting gravity" (Witten 1996), or "String theory incorporates and unifies the central principles of physics: quantum mechanics, gauge symmetry, and general relativity" (Polchinski 1998, p. 429). However, a more careful and more realistic analysis reveals that this is not really true: That string theory incorporates electrodynamics, and therefore vibration patterns of the type of massless spin-one photons, should be a minimum requirement for a theory, sometimes advertised as a 'theory of everything'. It is then not too surprising that it also allows for massless spin-two-excitations (gravitons) with their corresponding linear field equations. However, for the nonlinear back-coupling of the gravitational field energy, summing up to the full Einstein equations, string theory has never provided another recipe than the one indicated above as route 5, and which was established well before the different 'string revolutions' set in. So one has to come to the conclusion that string theory, besides lacking any successful experimental prediction or test in elementary particle physics, also does not really contribute to the cornerstones of general relativity, the equivalence principle, and the many spectacular predictions in astrophysics and cosmology, based on its nonlinearities and on the spacetime curvature (see Sect. 3.3).

The last route 6 in the MTW list, is based on a hypothesis due to Sakharov (1968), also referred to as the 'metric elasticity of space'. This suggests that the curvature of spacetime leads to a correction term to the vacuum polarization Lagrangian of elementary particles, which should (not too surprisingly) be proportional to the curvature scalar R. Equating the coupling constant of this term with the reciprocal of Newton's constant G results in a cutoff of the momentum integral at the Planck length $k_P \approx \sqrt{c^3/\hbar G} \approx 10^{-33}$ cm. To our knowledge this approach to general relativity has not led to any deeper understanding or indeed found any confirmation.

A very interesting and surprising 'derivation' of Einstein's field equations beyond the MTW list was propounded in Jacobson (1995). It is based on the consideration of black holes as thermodynamical systems fulfilling appropriate thermodynamic laws (Bardeen et al. 1973, Sect. 3.3). In this scheme, the heat Q is defined as the energy that flows across a causal (black hole or Rindler) horizon. The entropy S is chosen proportional to the horizon area, and the temperature T is chosen as the Unruh temperature (made up of vacuum fluctuations) of a uniformly accelerating observer hovering just inside the horizon. In order to apply the laws of equilibrium thermodynamics, the equivalence principle is invoked: one views a small neighborhood of each spacetime point p as a piece of flat spacetime, and chooses the null normal congruence, defining the horizon, such that its expansion and shear vanish at p. It can then be proven [with the help of the Raychaudhuri equation (Raychaudhuri 1955)] that the fundamental Clausius relation $dS = \delta Q/T$ of equilibrium thermodynamics can only be fulfilled if the spacetime curvature obeys Einstein's field equations, playing here the role of an equation of state. A subsequent paper (Eling et al. 2006) contains some corrections to these arguments, and a proposal for a generalization to nonequilibrium thermodynamics.

A last, quite formal route to Einstein's field equations of general relativity is closely connected with the four-dimensional analysis of Newtonian theory in Sects. 1.5 and 1.6 (Friedman 1983, Sects. V.1 and V.2). As mentioned there, a reformulation of Newtonian gravity as a four-dimensional spacetime theory provides a way to convert—with hindsight—the Newtonian field equations into their relativistic generalization by simply comparing the different spacetime structures of the two theories. [As Malament (2012, p. 162) puts it: "This seems to me one of the nicest routes to Einstein's equations."] In the four-dimensional Newtonian (or Galilean) theory we introduced a pair of symmetric and singular metric tensors on the spacetime manifold $\mathcal{M}$, viz., a temporal metric with components

$$t_{\alpha\beta} = t_{\alpha}t_{\beta} = \frac{\partial t}{\partial x^{\alpha}}\frac{\partial t}{\partial x^{\beta}} \ ,$$

and a spatial metric with components $s^{\alpha\beta}$. Then the formulation of Poisson's equation in Friedrichs (1927), viz.,

$$R_{\mu\nu} = t_{\mu\nu}s^{\sigma\kappa}\Phi_{;\sigma;\kappa} = 4\pi\varrho t_{\mu}t_{\nu} \ , \tag{3.3}$$

may be reformulated by introducing the tensor $T^{\mu\nu} := \varrho u^{\mu}u^{\nu}$, with $u^{\mu} = dx^{\mu}/dt$. This tensor field describes a continuous distribution of pressureless dust, with mass density $T^{00} = \varrho$ and momentum density $T^{\mu 0} = \varrho u^{\mu}$. Replacing the scalar field ϱ by this tensor $T_{\mu\nu} = t_{\mu\kappa}t_{\nu\sigma}T^{\kappa\sigma} = \varrho t_{\mu}t_{\nu}$ in (3.3), we get

$$R_{\mu\nu} = 4\pi\,T_{\mu\nu} \ . \tag{3.4}$$

In this way, the tensor field $T_{\mu\nu}$ is the source of Poisson's equation. Since we have $t_{\alpha\beta}T^{\alpha\beta} = \varrho$, (3.4) may be rewritten as

$$R_{\mu\nu} = 8\pi\left(t_{\mu\alpha}t_{\nu\beta}T^{\alpha\beta} - \frac{1}{2}t_{\mu\nu}t_{\alpha\beta}T^{\alpha\beta}\right) \ . \tag{3.5}$$

The *formal* transition from these Newtonian field equations to Einstein's field equations of general relativity is based on a change in the metric structure of space-time, i.e., by replacing the singular temporal metric $t_{\alpha\beta}$ and the singular spatial metric $s^{\alpha\beta}$ of Newtonian theory by the (non-singular) semi-Riemannnian metric tensor $g_{\mu\nu}$ of the curved space-time manifold in the final version of general relativity, and interpreting $T^{\mu\nu}$ now as the *total* energy–momentum tensor (replacing the mass density ϱ as the single source of gravity):

$$R_{\mu\nu} = 8\pi\left(g_{\mu\alpha}g_{\nu\beta}T^{\alpha\beta} - \frac{1}{2}g_{\mu\nu}g_{\alpha\beta}T^{\alpha\beta}\right) \ . \tag{3.6}$$

Since $R = 8\pi T$ and $T = g_{\alpha\beta}T^{\alpha\beta}$, this is equivalent to the usual formulation of Einstein's equation

$$G_{\mu\nu} = R_{\mu\nu} - \frac{1}{2}g_{\mu\nu}R = 8\pi T_{\mu\nu} .$$

In this way, the different (metric) structures of the spacetime manifold $\mathcal{M}$ underlying Newtonian theory and general relativity are the essential key to passing from the non-relativistic to the relativistic field equations. But in contrast to Einstein's field equation (3.6), in the Newtonian counterpart (3.5) neither a conservation law for the stress–energy tensor, nor the equation of motion are consequences of the field equations. As already remarked in Sect. 1.6, it is most satisfying that the reverse transition can also be made mathematically rigorous: Ehlers' frame theory offers the possibility to state such a limit relation between Newtonian and Einsteinian gravitation theory. (For the Newtonian limit of Einstein's theory of general relativity, see Ehlers (1981) and references therein.)

3.2   The Very Special Structure of Einstein's Field Equations and Some of Their Mathematical Consequences

The Einstein field equations $G_{\mu\nu}(g_{\lambda\kappa}) = R_{\mu\nu} - Rg_{\mu\nu}/2 = 8\pi T_{\mu\nu}$ are, for given energy–momentum tensor $T_{\mu\nu}$, a system of ten coupled partial differential equations (PDEs) of second order for the metric components $g_{\lambda\kappa}(x^\sigma)$. Since the second derivatives $g_{\mu\nu,\lambda\kappa}$ appear linearly in the Einstein tensor $G_{\mu\nu}$, they are (in the nomenclature of mathematics) quasilinear differential equations. But they are not semilinear, because the second derivatives have coefficients which depend on $g_{\mu\nu}$. The first derivatives $g_{\mu\nu,\lambda}$ appear quadratically, since $G_{\mu\nu}$ is quadratic in the Christoffel symbols $\Gamma^\mu_{\nu\lambda}$. The metric components $g_{\mu\nu}$ themselves enter the Einstein equations in a very complicated, highly nonlinear form, mainly because the dual tensor $g^{\mu\nu}$ also appears in the expression for $\Gamma^\mu_{\nu\lambda}$, and hence also in the Einstein tensor $G_{\mu\nu}$. It has to be said that the general mathematical theory of systems of quasilinear PDEs of second order is not much help in answering important physical questions in general relativity, mainly because the general theorems available are in a sense too weak and unspecific. Where such questions have been successfully answered, it was usually due to the very special structure of Einstein's field equations.

The ten Einstein field equations $G_{\mu\nu} = 8\pi T_{\mu\nu}$ are not all independent because, as we have seen already in Sect. 2.4, they are connected by the contracted Bianchi identities

$$G^{\mu\nu}{}_{;\nu} = 8\pi T^{\mu\nu}{}_{;\nu} = 0 . \tag{3.7}$$

Therefore, there are only six independent differential equations for the determination of the ten unknown metric potentials $g_{\mu\nu}(x^\lambda)$. The remaining four degrees of freedom correspond just to the freedom to perform coordinate transformations $x^\lambda \rightarrow x'^\lambda = f^\lambda(x^\rho)$ with four free functions f^λ. Equations (3.7) play a very characteristic and deep role in general relativity. This 'automatic conservation of energy and momentum' has the consequence that the equations of motion for any matter model already follow from Einstein's field equations. For this to be possible, the nonlinearity of Einstein's field equations is decisive. (In a linear theory like Maxwell's electrodynamics, or Newtonian physics, the superposition principle forbids such a consequence.)

For instance, if $T^{\mu\nu}$ describes a 'free point particle', we get the geodesic equation, as a generalization of Newton's first law of motion of Chap. 1. However, the details are quite delicate here, concerning the limit of infinitesimal spatial extent of the particles, the exclusion of a back-reaction of the particles on the 'exterior' spacetime, and the elimination of a possible inner rotation (spin) of the matter. Progress on these problems has only come quite recently in Ehlers and Geroch (2004) and Yang (2014). For the energy–momentum tensor of the electrodynamic field, (3.7) leads to the Coulomb and Lorentz force laws. If $T_{\mu\nu}$ describes a viscous fluid, there result the Navier–Stokes equations, and similarly for all other physically realistic matter models. In this way, the Einstein field equations encompass the whole of classical physics, and therefore are (by definition) the most universal, but also the most complicated equations of classical physics.

It is very satisfying that this fundamental relation (3.7) of general relativity also has a simple, purely geometrical visualization, as was first discovered in Cartan (1928), and explicated in more detail in Wheeler (1964) and (Misner et al. 1973, Chap. 15). The geometrical basis of this relation is the fact that the two-dimensional boundary of the three-dimensional boundary of any four-dimensional piece of a Riemannian manifold is identically zero. Application to Cartan's 'moment of rotation' $^*G = \mathbf{e}_\sigma G^{\sigma\tau}\mathrm{d}^3 S_\tau$ leads to the contracted Bianchi identities, and therefore, due to Einstein's field equations, to the automatic conservation of energy and momentum.

In any field of physics, the duty of its mathematical formulation is to predict the future evolution from some given 'initial' arrangement, as, e.g., in Newton's second law (see Sect. 1.6). It turns out that this program has a very special, quite difficult, but also interesting structure in general relativity. This begins with the fact that the choice of an 'initial hypersurface' $\mathscr{S}(x^0 = 0)$ is already a nontrivial problem, because $\mathscr{S}$ must be everywhere spacelike, already requiring some knowledge of the (to be derived) metric field $g_{\mu\nu}(x^\lambda)$. There are even solutions of Einstein's field equations, like the Gödel solution (Gödel 1949), which have no global, everywhere spacelike hypersurfaces whatsoever, and which even contain closed timelike curves. (Quite generally, the Gödel solution, although far from describing our actual cosmos, is a mathematically interesting counterexample for many physically expectations in general relativity. For example, in mathematical terms, the Gödel solution is not globally hyperbolic.)

On the other hand, locally, according to the equivalence principle, we have everywhere approximately Minkowski geometry, and therefore no problem with the definition of a time x^0 and a piece of a spacelike hypersurface $\mathscr{S}(x^0 = 0)$. But this quasi-local initial value or Cauchy problem still leads to severe problems in general relativity: which initial data can be freely prescribed on the (smooth) surface $\mathscr{S}$, and by which constraints can they be restricted? Do the Einstein equations provide a unique future evolution of these data, and what is the corresponding domain of dependence $D^+(\mathscr{S})$ which can be predicted? Relative to the time coordinate x^0 and the spatial coordinates x^i in $\mathscr{S}$ the ten Einstein equations can be divided into the following groups:

$$R_{00} = -\frac{1}{2} g^{ik} g_{ik,00} + M_{00} = 8\pi \left(T_{00} - \frac{1}{2} g_{00} T \right) , \tag{3.8}$$

$$R_{i0} = \frac{1}{2} g^{k0} g_{ik,00} + M_{i0} = 8\pi \left(T_{i0} - \frac{1}{2} g_{i0} T \right) , \tag{3.9}$$

$$R_{ik} = -\frac{1}{2} g^{00} g_{ik,00} + M_{ik} = 8\pi \left(T_{ik} - \frac{1}{2} g_{ik} T \right) , \tag{3.10}$$

where the terms $M_{\mu\nu}$ depend only on $g_{\mu\nu}$, $g_{\mu\nu,\lambda}$, $g_{\mu\nu,ik}$, and $g_{\mu\nu,i0}$, and not on second time derivatives, so that these can be prescribed (up to constraints) on $\mathscr{S}$. We see that there are no differential equations of second order in x^0 for the metric components $g_{\mu0}$. The derivatives $g_{\mu0,00}$ can therefore be freely attributed on $\mathscr{S}$. One could for instance choose $g_{\mu0,00} \equiv 0$ on $\mathscr{S}$. A more convenient choice are the so-called harmonic coordinates given (with the covariant derivative ∇_ν) by the condition $\nabla_\nu \nabla^\nu x^\mu = 0$, or by $\partial_\nu (\sqrt{-g} g^{\nu\mu}) = 0$, and which are comparable to the Lorenz gauge condition in electrodynamics. As already mentioned, the ten equations (3.8)–(3.10) are not all independent. Indeed, appropriate combinations give the following constraints which have to be satisfied on $\mathscr{S}$ (in geometrical terms, the constraints are the conditions that the hypersurface $\mathscr{S}$ can be embedded in some four-dimensional spacetime which is a solution of Einstein's equations):

$$g^{00} R_{i0} + g^{0k} R_{ik} = g^{00} M_{i0} + g^{0k} M_{ik} = 8\pi T_i^0 , \tag{3.11}$$

$$g^{00} R_{00} - g^{ik} R_{ik} = g^{00} M_{00} - g^{ik} M_{ik} = 16\pi T_0^0 , \tag{3.12}$$

which can also be combined to the Einstein equations $G_{\mu0} = 8\pi T_{\mu0}$. These constraints only have to be satisfied on $\mathscr{S}$, and are then (due to the contracted Bianchi identities) automatically satisfied at later times [in $D^+(\mathscr{S})$].

A detailed analysis of (3.11)–(3.12) shows that they are elliptic differential equations: the second space derivatives appear in the form $a^{ik} \partial^2 g_{\mu\nu}/\partial x^i \partial x^k$ with a positive definite matrix a^{ik}. For such equations, a well-posed problem is the boundary value problem (of Dirichlet or Neumann type). At least for a special case we will come back to such problems and their solvability below, and in Sect. 3.3.

The above analysis also answers the question of the number of (local) degrees of freedom of a general gravitational field. To begin with, we have on $\mathscr{S}$ the six quantities g_{ik} (the intrinsic curvature or first fundamental form of $\mathscr{S}$), and the six quantities $g_{ik,0}$ (the extrinsic curvature or second fundamental form of $\mathscr{S}$). However, due to the four constraints (3.11)–(3.12) and due to the additional coordinate freedom (gauge transformations) and freedom in the choice of the hypersurface $\mathscr{S}$, these reduce to *four* true degrees of freedom, of which two can be connected to the matter fields, and the other two to the vacuum gravitational field, e.g., to gravitational waves. (These results are quite similar to the situation in electrodynamics.) In general, it is of course impossible to reduce globally to four degrees of freedom, i.e., to eliminate the constraint equations explicitly. (Compare with the problem of anholonomic constraints in classical mechanics.)

Coming back to the initial value problem for the Einstein field equations, a first relatively simple observation was already made in Hilbert (2007): for analytic initial data, the Cauchy problem is uniquely solvable according to the well-known Cauchy–Kowalewski theorem. However, for general relativity this statement is of little use. A quite general argument for this is the fact that the Cauchy–Kowalewski theorem cannot deal with the issue of the causal propagation of any field. More concretely and specifically for gravity we may argue that any gravitational field should finally be produced by some matter, and surely we have no analyticity at the boundary between matter and vacuum. Furthermore, the physically interesting global vacuum solutions typically have singularities. For non-analytic initial data, but with differentiability properties such that the quantities in (3.8)–(3.12) are well defined, the analysis of the Cauchy problem affords specific, mathematically quite involved techniques.

After important preliminary work by Stellmacher, Lichnerowicz, and Leray, the breakthrough came with the seminal paper (Choquet-(Fourès-)Bruhat 1952) by Y. Choquet-Bruhat (see also Choquet-Bruhat 1962). Using harmonic coordinates and the techniques of Sobolev spaces, she succeeded in finding a successive solution of the problem, beginning with data sufficiently near flat spacetime. [For the mathematical details and for the proofs of the essential theorems, see the book Hawking and Ellis (1973, Chap. 7).] In this way, it was possible for the first time to prove the unique local solvability of the Cauchy problem, first of all for the vacuum Einstein equations. For the matter equations, appropriate conditions like energy conditions have to be fulfilled, and there are models (e.g., particles with spin greater than one) for which the initial value problem is not well-posed. In the later work (Choquet-Bruhat and Geroch 1969), it was shown that the Cauchy development covers a maximal region $D^+(\mathscr{S})$. (There are, however, solutions like the Gödel solution and the anti-DeSitter solution where globally different Cauchy surfaces have different maximal developments.)

The solution of the Cauchy problem proves not only the fundamental fact that general relativity is a predictive theory, it shows also that the gravitational field at an event $p \in \mathscr{M}$ depends only on initial data in and on the backward lightcone Γ_p^-, i.e., that gravitation propagates (locally) with the speed of light. Globally, however, it may happen that two events p and q are connected 'at the same time' by spacelike, lightlike, and timelike paths, or that there is more than one lightlike path between p and q, as for instance in the case of multiple images of quasars due to focusing by intervening matter (galaxies). For a recent textbook on the above and some other mathematical questions in general relativity, see Choquet-Bruhat (2008).

As announced above, the 'equivalent' to the Cauchy problem for hyperbolic (e.g., time evolution) equations is the boundary value problem (of Dirichlet or Neumann type) for elliptic (systems of) differential equations, as they show up in general relativity in the constraint equations (3.11) and (3.12), and for stationary systems with a global timelike Killing vector. As far as we are aware, there are so far no general existence and uniqueness results for such boundary value problems. In the case of the constraint equations, there are partial results, e.g., in O'Murchadha and York (1973). For the stationary vacuum Einstein equations, there is a general existence proof in Reula (1989) for asymptotically flat solutions, characterized by small data on a sphere, where it is also proven (using the implicit function theorem) that, for such data, the freedom is the same as that available from a multipole moment expansion at infinity. A more or less complete proof for the solvability of the Dirichlet boundary value problem is available in Schaudt and Pfister (1996) and Schaudt (1998) for the stationary and axisymmetric Einstein equations, which have particular importance for rotating equilibrium stars (see Sect. 3.3).

Of special interest here, is the fact that one cannot expect existence of such solutions for arbitrarily large data because it is qualitatively known that for data belonging to 'stars' with mass–radius ratios $M/R > 1$ and with angular momenta $J/R^2 > 1$, the gravitational collapse phenomenon and instability against rotational disruption forbid the existence of such solutions. (This is in characteristic contrast to the linear theory of electrodynamics, where the existence of boundary value solutions even for large data is no problem, but only the construction of explicit Green functions for appropriately simple geometries of the boundary.) Indeed in Schaudt and Pfister (1996) and Schaudt (1998), the existence of Dirichlet solutions was proven only for data restricted by $M/R < 1/100$ and $J/R^2 < 200$. For a topologically spherical boundary, one can always find coordinates such that the boundary is a geometrical sphere with radius R. The essential Einstein equations can then be written so that the second derivatives have the form of flat Laplacians in different space dimensions, and the exterior of the sphere is inverted to the interior. The Einstein equations can then be interpreted as a mapping in an appropriate Banach space, where the norm depends essentially on the derivatives of the metric potentials, and a more or less standard application of Banach's fixed point theorem proves by iteration the unique solvability of the Dirichlet problem for data limited as above. (Better optimized norms and more refined estimates may improve the above limits.) In principle, this method of proof is also applicable to the Einstein equations in the interior matter region of a rotating star, but the details then depend on the equation of state and the rotation law.

From the mathematical side, it is of particular interest that the quadratic dependence of the Einstein tensor on the metric derivatives is a kind of 'watershed' for elliptic systems of second-order PDEs: if these derivatives appear to sub-quadratic order, the Dirichlet problem is generically solvable (Ladyzhenskaya and

Uraltseva 1968); in the quadratic case additional conditions have to be fulfilled (as is obviously the case for the Einstein equations); if the derivatives appear to an order higher than quadratic, there are even analytic boundary data for which the Dirichlet problem is not solvable (Serrin 1969). [This is bad news for most alternative gravity theories like the $f(R)$ theories mentioned in Sect. 3.1.] It is remarkable that one of the best numerical codes (Bonazzola et al. 1993) for global solutions of rotating relativistic stars uses a mathematical scheme very similar to the above Dirichlet analysis, and the exponential convergence of this numerical iteration process is the typical property of a Banach fixed point argument. This gives hope that the above mathematical scheme may also work for the global (free boundary!) problem of rotating stars. However, as yet the remarks in Schaudt and Pfister (1996) about solutions for white dwarf stars and concerning the hoop conjecture (see Sect. 3.3) have only the character of plausibility arguments. A strict existence proof for rotating star solutions in general relativity seems still to be limited to the work (Heilig 1995) for slowly rotating and nearly Newtonian stars, where, due to the method of the implicit function theorem, nobody can say how relativistic these stars may be.

Another topic of fundamental physical importance in general relativity, but of considerable mathematical difficulty, is the so-called positive energy theorem. Quite generally, the concept of energy plays a very special and delicate role in general relativity. This begins with the fact that there is no notion of a local gravitational energy density: by analogy with electrodynamics and other relativistic field theories, such an energy density should be quadratic in the potential derivatives $g_{\mu\nu,\lambda}$. But according to the equivalence principle, these derivatives can all be made zero locally, and in appropriate coordinates. In more recent times there have been attempts to define so-called quasi-local energy expressions for finite regions of spacetime. However, all the different proposals (e.g., by Komar, Hawking, and Bartnik) have deficiencies, and they are in agreement and consistent only in cases of high symmetry. For an overview see Szabados (2004).

The positive energy theorem as such applies to spacelike slices of 'isolated systems', whose matter is concentrated in a finite region of space, and whose spatial metric approaches asymptotically the flat Euclidean metric. In appropriate coordinates one has for $r \to \infty$:

$$g_{ij} \to \delta_{ij} + O(r^{-1}), \quad g_{ij,k} = O(r^{-2}), \quad g_{ij,kl} = O(r^{-3}).$$

Strictly speaking, isolated systems represent an idealization, because all physical systems are finally a part of our whole cosmos. However, for systems like planets, stars, and even whole galaxies, the influence of the rest of the universe is in most cases almost arbitrarily small. Furthermore, for the cosmos as a whole the concept of energy loses much of its usual significance anyway, since we cannot define or measure the total energy 'from the outside'. And there are surely also problems with the usual energy conservation law because, due to phenomena like the initial big bang and cosmic expansion, there is no time symmetry, so the usual connection between time symmetry and conservation of energy—according to Noether's theorem—does not apply.

Coming back to isolated systems, it is qualitatively expected that a total energy can be consistently defined for them, and that this energy will be non-negative if the matter obeys physically reasonable local energy conditions. The standard condition of this type is the 'dominant energy condition' $T^{00} \geq |T^{\mu\nu}|$ for all μ, ν. (For quantum matter and vacuum polarization, this condition may be violated. Also in cosmology other matter forms may play a role.) In terms of Newtonian gravity and special relativity, the problem with the positivity of energy is the following: whereas the mass energy given by $E = mc^2$ and the kinetic energy are surely positive, the gravitational interaction energy is, due to the attractive gravitational force, always negative, and the question is whether the sum of all these energy forms will nevertheless be non-negative in all physically realistic cases of isolated systems.

The answer to this question in the universal classical theory of general relativity turned out to be conceptually and mathematically quite difficult. The usual definition of total energy of a spacelike slice with asymptotically Euclidean metric is given by the so-called ADM energy

$$E = P^0 = \lim_{r \to \infty} \frac{1}{16\pi} \int dS_i \, (\partial_j g_{ij} - \partial_i g_{jj}) \,,$$

according to the Hamiltonian formulation of general relativity by Arnowitt et al. (1962). The corresponding three-momentum is given by

$$P^i = \lim_{r \to \infty} \frac{1}{8\pi} \int dS_j \, (K^i_j - \delta^i_j K) \,,$$

with the extrinsic curvature tensor K_{ij} of the spacelike slice. After obtaining some results for the positivity of this energy P^0 in special cases, the first more general problem was tackled in Brill and Deser (1968): the Minkowski spacetime (with $P^0 = 0$) is a 'local' minimum of energy, i.e., all 'small' deviations from Minkowski have positive energy. Some mathematical deficiencies of this work were removed in Choquet-Bruhat and Marsden (1976). The full proof of the positive energy conjecture, i.e., that Minkowski is the absolute energy minimum for all isolated systems with dominant energy condition (or, more generally, that P^μ is timelike and future-directed, and vanishes only for flat spacetime) was then spelt out in two seminal papers by Schoen and Yau, first in Schoen and Yau (1979) for the case where P^0 is calculated on a maximal slice (with $K = 0$), then in Schoen and Yau (1981) for the general case, where the spacelike slice was also allowed to have more than one asymptotically flat end.

These papers were based on a proof by contradiction within a variational analysis, and they had to use quite heavy mathematical machinery (geometric measure theory and Sobolev inequalities). Surprisingly, shortly afterwards a completely different (and in a way much simpler) proof was presented in Witten (1981), motivated by attempts to build a so-called supergravity, and using spinor methods.

A loophole in this proof (existence of solutions of the Witten equation) was independently closed shortly afterwards in Taubes and Parker (1982) and Reula (1982).

Whereas all other classical physical theories leave open the 'absolute zero' of energy, general relativity is now singled out by fixing this value due to the positive energy theorem. In any case, this theorem represents a deep self-consistency of general relativity. And this fact and the methods of proof have turned out to be important also for other fundamental questions in general relativity.

Besides the above ADM energy, another energy expression for isolated systems is of importance in general relativity, particularly in connection with gravitational waves. This 'Bondi energy' is not defined at spacelike infinity but at lightlike infinity (on Scri $\mathscr{J}$), and this energy is also expected to be non-negative for 'reasonable' physical systems, and typically decreasing in time due to the emission of gravitational radiation. Indeed, these conjectures were proven at the same time, simultaneously by geometric methods in Schoen and Yau (1982) and by spinor methods in Horowitz and Perry (1982) and Ludvigsen and Vickers (1982).

Concerning the positive energy theorem, most alternative gravity theories are inferior to general relativity in the sense that the positivity question remains open, or that it is explicitly proven that the theorem is wrong. A prominent example of the latter case is the five-dimensional Kaluza–Klein theory which was originally (in the years 1921–1926) constructed as a unification of gravity with electromagnetism, but which has, in more recent times, received attention and extensions far beyond this aim. However, it was shown in Witten (1982) that the ground state of this theory is unstable against a semiclassical decay process. And in Brill and Pfister (1989), relatively simple (although topologically nontrivial) explicit, asymptotically flat solutions were constructed which have arbitrarily negative energy. Moreover, in string and superstring theories, no positive energy theorem is generally valid, as shown, e.g., in Brill and Horowitz (1991). Quite generally, there is reason to be very critical of the higher-dimensional theories which have been so popular in recent times: these higher dimensions provide so many new 'degrees of freedom', and leave so many alternatives undecided, that the often unequivocal physical consequences in four-dimensional spacetime get completely lost. A useful intuitive view of this dilemma opens if one goes in the other direction and imagines living in a two-dimensional space, with so few structures, without bridges, tunnels, rivers, trees, birds, airplanes, etc. [See also the popular book Square (1884).]

With the proof of the positive energy theorem, and therefore with the fact that Minkowski spacetime is the ground state for isolated systems in general relativity, the question arises—as with the ground state in any other field of theoretical physics—whether this solution is stable with respect to arbitrary (small) modifications, because in nature we can never expect to have the idealization of strict vacuum and maximal symmetry, as is the case for the Minkowski solution. And by stability is meant not only stability against first order perturbations, but full so-called nonlinear stability. In general relativity it turned out that this proof (usually confined to the initial data set, i.e., to the Cauchy problem) is mathematically extremely fastidious and involved. The first proof in Christodoulou and Klainerman (1993) filled a book of more than 500 pages, although it was later shortened to an article Klainerman and Nicolò (1999) of 85 pages.

Besides the Minkowski solution, the same stability question was also addressed for the vacuum and maximally symmetric solutions of the Einstein equations with cosmological constant Λ, i.e., where the Einstein tensor $G_{\mu\nu}$ is replaced by $G_{\mu\nu} + \Lambda g_{\mu\nu}$. For the so-called deSitter solution with $\Lambda > 0$, it turned out that the proof in Friedrich (1986) is simpler than for the Minkowski solution. In contrast, for the anti-deSitter solution with $\Lambda < 0$ (which is not globally hyperbolic!), the stability question is still not completely decided. There are indications of unstable modes in Bizoń and Rostworowski (2011), but it is still unclear how large and realistic such a class of unstable modes is, and how it depends on boundary conditions (Friedrich 2014).

Among the many additional mathematical consequences of the very special structure of Einstein's field equations, the last one we would like to address explicitly is possibly the most surprising and the most dramatic, namely the so-called singularity theorems. Already at the birth of general relativity, the Schwarzschild solution

$$ds^2 = -(1 - 2M/r)dt^2 + (1 - 2M/r)^{-1}dr^2 + r^2 d\Omega^2$$

revealed that there are exact and physically realistic solutions of Einstein's field equations which show space-time singularities: besides the real (curvature) singularity at the center $r = 0$, it was thought for a long time that the region $r = 2M$ is also singular. Only around the year 1960 was it proven that this region is only a coordinate singularity, although nevertheless a remarkable region called a horizon from whose interior $r < 2M$ no signals can escape to the exterior $r > 2M$. Some authors, e.g., Lifschitz and Khalatnikov (1963), believed that such singularities are only due to the high symmetry of the Schwarzschild solution, but then in the pioneering paper (Penrose 1965), R. Penrose showed that the appearance of singularities is a generic and realistic phenomenon in general relativity. To some extent, this theory predicts its own limits!

As a central concept in his first proof of a singularity theorem, he introduced the so-called trapped surface, a closed, spacelike two-surface $\mathscr{T}^2$ with the property that the two systems of null geodesics which meet $\mathscr{T}^2$ orthogonally both converge locally to the future. While 'normally' one of these systems always diverges, in the stellar collapse process, such trapped surfaces typically appear due to the extremely strong gravitational attraction, and this independently of any symmetry assumptions. Penrose could then prove the following: if a connected, globally hyperbolic spacetime with a noncompact Cauchy surface $\mathscr{S}$ fulfils the energy condition $R_{\mu\nu}k^\mu k^\nu \geq 0$ for all null vectors k^μ, and if it contains a trapped surface $\mathscr{T}^2$, it has at least one inextendible future directed null-geodesic orthogonal to $\mathscr{T}^2$ which has affine length no greater than $2/|\Theta|$, where $\Theta < 0$ denotes the maximum value of the 'expansion' of all systems of null geodesics orthogonal to $\mathscr{T}^2$. [For a proof of this and the following singularity theorems see Hawking and Ellis (1973, Sect. 8.2).] The main 'weak point' of the above theorem is the demand for a Cauchy surface $\mathscr{S}$.

Further important work, mainly by Hawking, culminated in the singularity theorem (Hawking and Penrose 1970), which can be stated as follows. A spacetime $(\mathcal{M}, g_{\mu\nu})$ is not timelike and null geodesically complete if:

- The strong energy condition $R_{\mu\nu}k^\mu k^\nu \geq 0$ is valid for all timelike and null vectors k^μ.
- A 'timelike generic condition' is valid, viz., each timelike geodesic with tangent vector k^μ has at least one point with $R_{\mu\nu\lambda\kappa}k^\mu k^\nu \neq 0$.
- There are no closed timelike curves.
- At least one of the following three conditions holds:

  - $(\mathcal{M}, g_{\mu\nu})$ describes a closed universe.
  - $(\mathcal{M}, g_{\mu\nu})$ possesses a trapped surface.
  - There exists an event $p \in \mathcal{M}$ such that the expansion Θ of all future- or past-directed null geodesics emanating from p becomes negative, i.e., the null geodesics from p are focused by matter or curvature.

This theorem is considerably more general than the original Penrose theorem. In particular it gives good reason to believe that our universe is singular. Although the theorems do not provide strict mathematical arguments concerning the character of the singularities, it is quite plausible that, for the concrete astrophysical phenomena represented by a star collapse or the cosmological Big Bang, the spacetime curvature diverges. And in contrast to other divergencies which have appeared in the history of theoretical physics, e.g., the 'resonance catastrophe' for a forced oscillator which is cured by including a damping term, there is no obvious correcting mechanism for the singularities of general relativity, at least in classical physics. [There are cosmological models, violating the strong energy condition, and thereby avoiding the Big Bang singularity, e.g., Rose (1986), but such models are far from solving all singularity problems in general relativity.]

There are indications that quantum effects may change some of the above unpleasant consequences of classical general relativity, e.g., Hawking radiation (Hawking 1975) at the horizon of black holes, or the mitigation of the Big Bang singularity by effects of loop quantum gravity (see, e.g., Bojowald 2001). But since, even after decades of intensive research, no consistent union of general relativity and quantum physics is yet visible, we encounter here the deepest crisis for general relativity, and indeed for the whole of theoretical physics.

As a last topic in the realm of singularities in general relativity, we briefly mention the 'cosmic censorship hypothesis' by Penrose (1969): nearly all singularities of general relativity are 'hidden' by event horizons, and no 'naked' singularities other than the Big Bang exist in our universe. Some sort of hypothesis of this kind is of central importance for physics, because naked singularities would destroy the predictive power of physics (in the 'future' of these singularities). However, till now no mathematically clear-cut and physically realistic theorem of this type has been found, and there are even (somewhat artificial) counterexamples, e.g., Shapiro and Teukolsky (1991).

3.3   Special, Important, and Sometimes Spectacular Predictions of General Relativity in Physics, Astrophysics, and Cosmology

Surely the most spectacular prediction of general relativity is the phenomenon of black holes (BH). And even more spectacular is the fact that these in every respect extreme objects are not just artefacts of our theoretical models but obviously abound in the universe as superheavy galactic nuclei with more than 10^9 solar masses, as collapsed heavy stars of a few solar masses, and as 'midweight' black holes. [One is here reminded of a remark which Einstein made in another connection, namely in a letter to H. Weyl (Schulman 1998, Doc. 551), dated to 31 May 1918, concerning Weyl's gauge principle: "Could one really accuse the Lord of inconsequence, if He missed the opportunity to harmonize the physical world in the way found by you?"] The simplest example of a black hole is in principle already given by the Schwarzschild solution from 1916. But, as already mentioned in the previous section, Schwarzschild, Einstein, and the other physicists of that time dismissed the 'singularities' at $r = 2M$ and $r = 0$ as unrealistic mathematical artefacts.

A first hint that astrophysical objects may show extreme gravitational properties came from the work of Chandrasekhar. He considered static, spherical, and cold white dwarf stars in which the quantum mechanical degeneracy pressure of electrons (due to the Pauli exclusion principle) is in equilibrium with the gravitational pressure (Chandrasekhar 1931a,b). He found that the degeneracy pressure can change from the nonrelativistic behavior $p_d \sim n^{5/3}$ to the relativistic behavior $p_d \sim n^{4/3}$, similar to the gravitational pressure. Therefore, the particle density n drops out, and there results a limiting value for the so-called Chandrasekhar mass $M_C = (c\hbar/G)^{3/2}/m_N^2 \approx 1.7 M_\odot$, where m_N is the nucleon mass and $M_\odot$ is the solar mass. (It is quite remarkable that the fundamental constants $c, \hbar, G$ together with the elementary particle quantity m_N define in this way an important mass limit for global astrophysical objects!) Since in the year 1931 the neutron had not yet been discovered, and therefore inverse β-decay was not yet known, Chandrasekhar could not have known that, in a heavy burnt out star, the electron gas together with the nuclei could transform into a neutron star. But in the case of relativistic neutrons, the above consequence for a limiting mass M_C persists essentially unchanged, although with the difference that the mass density can now reach values $\varrho \approx 10^{16} \, \text{g/cm}^3$, beyond the nuclear density, and the mass/radius ratio can come near to the Schwarzschild value 1/2, i.e., general relativistic effects become important, and for the first time the full nonlinearity of general relativity should be relevant here. (Quite generally, for the special mathematical consequences in the previous section, and for the special physical predictions in this section, the nonlinearity of Einstein's field equations is crucial.)

In any case, such a mass limit, i.e., a limit for an equilibrium state in a heavy star, raises the question of what happens to a star which, after burning all its nuclear fuel, still has a mass greater than M_C. The first quite general analysis of static spherical cold stars within the framework of general relativity was carried out in Oppenheimer and Volkoff (1939) and Oppenheimer and Snyder (1939). For instance, they derived the following differential equation governing the dependence of the pressure p on the radius r:

$$\frac{dp}{dr} = -\frac{(\varrho + p)[m(r) + 4\pi r^3 p]}{r[r - 2m(r)]} , \qquad (3.13)$$

where $m(r)$ is the mass inside the sphere with radius r. [In the literature (3.13) is often called the Tolman–Oppenheimer–Volkoff (TOV) equation, although this general equation does not appear in Tolman (1939), but only some of its solutions for special equations of state.] Compared to the Newtonian equation $dp/dr = -\varrho m(r)/r^2$, the numerator of (3.13) gets bigger due to the pressure contributions, and the denominator gets smaller. Therefore, in a type of 'avalanche effect', the pressure can increase dramatically on the way to the stellar center. (In contrast to standard thermodynamics, in strong field general relativity, the pressure acts not as a repulsive expansion effect, but contributes to contraction!) A particular consequence of this is that, for a (hypothetical) star with constant mass density ϱ (interior Schwarzschild solution), the pressure already diverges at $r = (9/8)2M$, i.e., before reaching the 'horizon value' $r = 2M$. Oppenheimer and Snyder also anticipated the essential property of these objects, which later became known as black holes (Oppenheimer and Snyder 1939): "The star thus tends to close itself off from any communication with a distant observer." [In the review article (Israel 1987), this paper is called "the most daring and uncannily prophetic paper ever published in the field".]

As mentioned in Sect. 3.2, a global analysis of the (vacuum) Schwarzschild solution was first performed only around the year 1960. It was found that the maximal extension of this geometry has two flat ends, and that a so-called horizon appears at $r = 2M$ as a type of a one-way membrane through which matter, light, and any kind of information can only proceed from the exterior $r > 2M$ to the interior $r < 2M$, but never vice versa. A decisive push, the beginning of a more general 'black hole physics', was triggered by the discovery of the Kerr metric. However, as already mentioned in Sect. 2.2, this metric was found not so much as the result of a systematic search for solutions of rotating stars or black holes, but by lucky circumstances in a mathematical study of Petrov type II vacuum solutions of Einstein's field equations. It should also be said that this solution represents presumably 'only' a rotating black hole, and not (as sometimes stated in the literature) the exterior of non-collapsed rotating stars: intensive research over several decades has never produced a so-called 'interior Kerr metric' for a physically realistic equation of state. Moreover, the analysis of the solution manifolds for the Dirichlet problem in the exterior and interior of a rotating star in Schaudt and Pfister (1996) makes it entirely plausible that such a solution does not actually exist. And in contrast to the spherical case where the Birkhoff theorem guarantees the uniqueness (and staticity!) of the vacuum Schwarzschild solution, in the case of stationary rotation, a constructive solution generation technique is available (Stephani et al. 2003), producing a huge solution manifold of vacuum solutions of the field equations, depending on two free functions of one variable (e.g., the axis potentials). For a prominent interior matter solution, the Wahlquist solution, it is even explicitly proven in Bradley et al. (2000) that there cannot be a smooth join to any asymptotically flat exterior vacuum solution, and in particular not to the Kerr metric.

The growing theoretical activity in the study of extreme gravitational objects like black holes coincided with and was supported by new and dramatic astrophysical discoveries. Around the year 1963, so-called quasars were observed. These are very distant but extremely luminous objects, for which masses of the order of 10^9 solar masses in relatively small volumes were predicted, and which were connected with galactic nuclei, and possibly with black holes (Lynden-Bell 1969), a name coined by Wheeler (1968). In 1967 the first pulsar was discovered (Hewish et al. 1968) and interpreted as a rapidly rotating neutron star, presumably the end product of a supernova explosion. These and several other relevant observations inaugurated the new field of relativistic astrophysics.

Of special importance was the discovery of the pulsar PSR 1913+16 in a binary system (Hulse and Taylor 1975). This provided ways to test many predictions of general relativity with previously unattainable precision. Observation of this system over a long time resulted in the measurement of a decrease in its orbital period (Taylor et al. 1979), for which the only natural explanation was an energy loss due to the emission of gravitational radiation, and this in numerical agreement with Einstein's quadrupole formula from Einstein (1918). In more recent times, a binary system consisting of two pulsars has been discovered, providing a unique laboratory for strong gravity fields, and this has confirmed many predictions of general relativity with unprecedented precision (Kramer and Wex 2009).

Of course, by definition, black holes can never be observed directly, and so far it has not been possible to probe the scale of an event horizon of any BH candidate. The clearest hint as to the existence of such extreme objects comes from measuring the orbits of stars surrounding the object. For instance, by applying essentially only Kepler's laws, it has been found (Genzel et al. 2010) that, in the center of the Milky Way (in Sagittarius A*, and within a central region of parsec diameter), there resides a massive BH of about $4.4 \times 10^6 M_\odot$. Likewise for the processes in the nuclei of most other galaxies, the only natural explanation is the existence of BHs, some of them with masses greater than $10^9 M_\odot$ (McConnell et al. 2011). Besides these supermassive BHs, there are reliable astrophysical observations of many 'midweight BHs' with masses between $10^3 M_\odot$ and $10^6 M_\odot$ (Greene and Ho 2004). Furthermore, just in our own galaxy, there exist millions of 'small BHs' with masses between $5 M_\odot$ and $30 M_\odot$ (Narayan and McClintock 2014), which are presumably the end product of stellar collapse.

Most of the observed BHs mentioned above are of course not isolated, stationary systems describable by the vacuum Schwarzschild or Kerr solution. Rather they are characterized (and in most cases only 'visible') by highly dynamical processes like the infall of gas and matter, the formation of fast rotating accretion discs, and/or the evolution of gamma ray bursts, the most energetic transient phenomena in the universe.

A general theoretical analysis of BH solutions in general relativity began around the year 1967 and concentrated first on static and stationary systems, which can be seen as the equilibrium end products, once the highly dynamical processes around a star collapse, mentioned above, have settled down. Some truly remarkable results are the black hole uniqueness theorems, also called the no-hair hair theorem, which were first derived (not in complete mathematical generality and rigor) by W. Israel, B. Carter, S. Hawking, and D.C. Robinson, with later generalizations and improvements, usually by applying the positive energy theorem (see Sect. 3.2). Today quite complete and useful reviews are available in Heusler (1996) and Chruściel et al. (2012).

The theorems say essentially that a static BH with horizon is necessarily spherical and identical with the Schwarzschild BH, a stationary BH rotates necessarily axisymmetrically, and is identical with the Kerr metric, a static, charged BH is uniquely represented by the Reissner–Nordström metric, and a rotating, charged BH is represented by the so-called Kerr–Newman metric (Newman et al. 1965). Therefore the stationary BH solutions of general relativity are quite simple, one might even say the most elementary physical systems, being completely characterized by the three parameters of mass M, angular momentum J, and charge Q. The latter does not usually play a role in astrophysics, but is of interest in theory, mainly due to the fact that the coupled Einstein–Maxwell equations are structurally quite similar to the pure Einstein vacuum equations, in particular in the stationary case.

Of course, the above theorems also say that all higher mass multipoles, angular momenta, and electromagnetic multipoles are uniquely determined by M, J, and Q. It should also be noted that this fact once again singles out general relativity and electromagnetism, because in alternative or higher-dimensional gravity theories, and by coupling to other fields like Yang–Mills, additional 'hairs' for BH solutions typically appear. Another 'miraculous' property of general relativity shows up if one considers the gyromagnetic ratio $g = 2Mm/QJ$ (where m is the magnetic moment, connected with J and Q) of the above stationary BH solutions. As first observed in Carter (1968), all these solutions have the value $g = 2$. Actually, this fact comprises two 'miracles':

- g has the same value for all rotating, charged BHs, instead of being some nontrivial function of the two dimensionless quantities J/M^2 and Q/M.
- The g-value for the 'elementary' BHs of general relativity coincides with the value $g = 2$ of the most elementary quantum particles like the electron, the muon (neglecting radiation corrections), and even the W-meson, for which S. Weinberg predicted this property in Weinberg (1970), long before the existence of this exchange particle was confirmed experimentally in 1983.

It is still an open question whether this coincidence signals a deeper common root between general relativity and the quantum theory of elementary particle physics, which may even assist in the task of unifying these experimentally extremely successful theories.

In Pfister and King (2003) we have summarized in more detail some of these remarkable properties of the g-factor in electrodynamics, quantum theory, and general relativity, and we have shown that a value $g \approx 2$ shows extreme 'robustness' in a class of rotating, charged, high-mass shell models. Besides the stationary BH solutions, a value $g = 2$ is also valid for a huge class of other electro-vacuum solutions of general relativity, e.g., for the Tomimatsu–Sato class, and for all solutions generated by a Harrison transformation from an arbitrary (uncharged) stationary and axisymmetric vacuum solution of the Einstein equations, i.e., for a manifold of solutions depending on two arbitrary real functions and one complex parameter (Klein 2002). In contrast, in generalized gravity theories such as Kaluza–Klein, supergravity, theories with dilatons, string theories, etc., a value $g = 2$ is not equally preferred, mainly because there is no comparable no-hair theorem, and additional new (fermionic, dilaton, etc.) hairs contribute to the angular momentum and to the g-factor. [For the relevant references see Pfister and King (2003).]

The fact that the stationary, i.e., the equilibrium BH solutions of general relativity are completely characterized by the three quantities M, J, and Q, is reminiscent of the fact that thermodynamical systems in equilibrium are also determined by very few variables (potentials), in contrast to their complicated and multi-faceted non-equilibrium counterparts. Indeed, the analogy between BH physics and thermodynamics goes much deeper. As first presented in detail in Bardeen et al. (1973), there are "four laws of black hole mechanics" [see also Wald (1984, Sects. 12.5 and 14.4) and Heusler (1996, Chap. 7)]. The zeroth law says that the surface gravity κ is constant over the whole horizon surface of any equilibrium BH, by analogy with the fact that an equilibrium thermodynamic system has uniform temperature. The first law reads in the vacuum case $\delta M = \kappa \delta A / 8\pi + \Omega_H \delta J_H$, where A is the horizon surface and Ω_H is its (constant!) angular velocity, and is the analogue of the thermodynamic energy variation law $\delta E = T\delta S + p\delta V$. The reduction of the surface gravity with angular momentum (for constant M) implied by the first law can be thought of as a centrifugal effect. If the BH is surrounded by stationary matter, e.g., in the form of rotating rings, there are generalizations of this law with additional work terms.

The analogy between horizon area A and entropy S, already indicated by the first law, is much strengthened by the second law: the horizon area of a BH can never decrease with time (Hawking 1971). The third law of BH mechanics says that it is impossible to reduce the surface gravity κ to zero, by analogy with the fact that it is impossible to reach absolute zero temperature $T = 0$ for a thermodynamic system. (However, there is no analogue of the thermodynamic law $\lim_{T\to 0} S = 0$, because the horizon area A can remain finite as $\kappa \to 0$.)

In a way the deepest, but also the most controversial of the four laws of BH mechanics is the second. An unsolved question is whether the horizon area A is only analogous to the entropy, or whether it can be seen as a real entropy contribution in all respects, as, for instance, Bekenstein (1973) has postulated with his 'generalized entropy'

$$S_B = S_{\text{matter}} + \frac{1}{4}\frac{kc^3}{G\hbar}A \,, \tag{3.14}$$

and a generalized second law $\delta S_B \geq 0$. By analogy with the statistical mechanical foundation of thermodynamics, this would require the horizon area A to be understood as the logarithm of the number of some microscopic, presumably quantum mechanical states. In Wald (1984, p. 418), this difficulty is characterized by the statement: "Since the nature of time in general relativity is drastically different from that in nongravitational physics, it is not clear precisely how the generalized second law will arise, even if $A/4$ is a measure of the number of internal states of a black hole." A connection between the elementary BH states of general relativity and quantum theory is strengthened by the phenomenon of the (quantum-mechanical) Hawking radiation (Hawking 1975) with temperature proportional to the reciprocal of the BH mass, but the physical basis for such a connection remains unclear, and will presumably continue to do so until a consistent quantum theory of gravitation is available. A particular, much debated problem in this connection is the information loss when matter and radiation are irredeemably swallowed by a BH, in apparent or real violation of the unitary time evolution of quantum mechanics. That the four laws of BH mechanics represent a central phenomenon of general relativity is also evident from the fact, already explained in Sect. 3.1, that these thermodynamic properties provide a new route to Einstein's field equations as an equation of state (Jacobson 1995).

The above laws of BH mechanics are, as indicated, not only valid for the pure vacuum BHs, but also, possibly with appropriate modifications, for BHs surrounded by stationary matter. Another group of properties (equalities and inequalities) of the pure BHs which extend to the more realistic BH + matter systems, has been analyzed only relatively recently. It began with the observation (Ansorg and Pfister 2008) that the relation $4J^2 + Q^4 = (A/4\pi)^2$ for the degenerate Kerr–Newman BH solutions (with surface gravity $\kappa = 0$) is also valid if the BH is surrounded by quite arbitrary stationary, axisymmetric, and equatorially symmetric matter. (The solutions do not even have to be asymptotically flat!) On the basis of a large number of numerical solutions, it was also conjectured that the inequality

$$4J^2 + Q^4 \leq (A/4\pi)^2 \, , \tag{3.15}$$

valid for all non-degenerate Kerr–Newman BHs (with equality only in the degenerate case), extends to the case with surrounding matter. This conjecture was then proven in Hennig et al. (2008, 2010). That the inequality (3.15) extends even to dynamical (but still axisymmetric) systems was first conjectured in Dain (2010), and proven (for the uncharged case) in Dain and Reiris (2011). For the proof of the charged case see Gabach Clément and Jaramillo (2012) and Gabach Clément et al. (2013). (This inequality has even been extended to higher dimensions, and, e.g., to Einstein–Maxwell–dilaton gravity.)

Another interesting inequality between the parameters appearing in the Kerr (or Kerr–Newman) metric was also recently proven [see Dain (2008), and references therein]: for all axisymmetric, vacuum (or electro-vacuum), asymptotically flat (but not necessarily stationary!), and maximal initial data (the extrinsic curvature is trace-free) with two asymptotic ends one has $|J| \leq M^2$, with equality only for the extreme Kerr solution. In a quasi-Newtonian picture one can interpret this inequality as follows: in a collapse, the gravitational attraction ($\approx M^2/r^2$) at the horizon ($r \approx M$) dominates over the centrifugal repulsive force ($\approx J^2/Mr^3$). The electromagnetic extension was here proven in Chrusciel and Lopes Costa (2009).

A last, even deeper inequality between the parameters of a BH is the 'Penrose inequality'

$$A/16\pi \leq M^2 \qquad (3.16)$$

(with equality only for the Schwarzschild spacetime), which was conjectured in Penrose (1973). It was motivated at least partly by the cosmic censorship conjecture, as mentioned in Sect. 3.2, because a counterexample to this inequality would imply that cosmic censorship is not true. Since this inequality does not contain the angular momentum J explicitly, one can expect it to hold without any symmetry restrictions. Of course, the Penrose inequality can be regarded as a strengthening of the positive mass theorem (see Sect. 3.2). Up to now it has only been possible to prove the so-called Riemannian Penrose inequality (with time-symmetric initial data with extrinsic curvature $K_{ab} \equiv 0$), but this with two quite different methods [Huisken and Ilmanen (2001) using the inverse mean curvature flow, and Bray (2001) using the positive mass theorem]. Moreover, both of these methods differ from those used to show the other inequalities mentioned above. Concerning the general (non-time-symmetric) Penrose inequality, the extension to rotating systems with black holes would of course be of particular astrophysical interest. For a very recent, excellent, and non-technical review of all the above 'geometric inequalities' (reminiscent of the well-known isoperimetric inequality $4\pi A \leq L^2$ between the length L of a plane closed curve and its enclosed area A, with equality only for the circle), see Dain (2014a).

Important and of astrophysical relevance is the possible extension of the above inequalities between mass M, angular momentum J, and surface area A from black holes to realistic, non-collapsed bodies, e.g., stars. A first, quite famous inequality conjecture for the borderline between these two cases is the so-called hoop conjecture in Thorne (1972): horizons form when and only when a mass M gets compacted into a region whose circumference in *every* direction is $\mathscr{C} \lesssim 4\pi GM/c^2$. Added is here the nice and realistic proviso: "Like most conjectures, this one is sufficiently vague to leave room for many different mathematical formulations." Indeed, the difficult and partly controversial topics in attempts to reach at least partial proofs of this conjecture concern the definition of the circumference $\mathscr{C}$ (or of some other measure for the extension of the body) and of the mass M, whether it be the asymptotic ADM mass or one of the (many different) quasi-local mass proposals. Generally, one can say that, in the spherical case, the conjecture is more or less settled, but the non-spherical, e.g., the fast rotating case is still wide open.

Where strict mathematical theorems are available, the conditions are usually far from astrophysical reality, and most of the positive answers address only the 'when' part of the conjecture. However, careful numerical analyses make it plausible that some form of this conjecture is true also in the general case. For a useful review of the hoop conjecture (up to the year 1991) see Flanagan (1991).

Very recently, a universal inequality $\mathscr{R}^2(U) \gtrsim |J(U)|$ has been conjectured in Dain (2014b), relating the 'size' $\mathscr{R}$ and angular momentum J for all (non-collapsed) bodies U, where $\gtrsim$ denotes only an order of magnitude, depending on the definition of $\mathscr{R}(U)$. In the case of a maximal, axially symmetric initial data set with constant energy density in U, fulfilling the dominant energy condition, a precise definition of $\mathscr{R}(U)$ and a complete proof of the conjecture have been provided. Remarkably, the above inequality of general relativity, if applied to elementary particles with spin s, results in a size $\mathscr{R}_0 \approx [s(s+1)]^{1/4} l_P$, with the Planck length $l_P = (G\hbar/c^3)^{1/2}$.

An interesting mathematical phenomenon in the collapse of matter to a BH has been discovered and carefully studied only relatively recently, namely so-called critical phenomena occurring during this gravitational collapse. [Compare the general theory of critical phenomena, appearing, e.g., in phase transitions, and in nonlinear dynamical systems, as described in Sornette (2009).] The example in which this gravitational phenomenon was first (numerically) derived is the simplest nontrivial matter model one can think of, viz., the spherically symmetric collapse of a massless scalar field, for which deep analytic studies had already been carried out, e.g., in Christodoulou (1991). In Choptuik (1993), the following new properties were found (with extremely high numerical precision): in the dependence on a parameter p which characterizes the strength of the gravitational self-interaction of the scalar field, the BH limit p^* has the character of a universal, self-similar attractor, and the BH mass scales according to $M_{BH} \sim |p - p^*|^\gamma$, with critical exponent $\gamma \approx 0.37$. (In this way, arbitrarily small BH masses, in comparison to the overall ADM mass of the system, can be created.) Whether and in which way these results transfer to more realistic matter models and to non-spherical, e.g., rotating systems, has not yet been sufficiently worked out, but some numerical studies give reason to believe that this critical collapse is of a more general nature in the nonlinear theory general relativity [see the review Gundlach and Martín-García (2007)].

This brings us to the question of existence and properties of general non-collapsed bodies in general relativity. If we are guided by astrophysical observations, a ubiquitous and relatively simple class is represented by the extremely long-lived stars which can be well approximated by an ideal fluid in stationary and axisymmetric rotation. (Even for brightly shining stars, the energy loss by radiation is extremely small compared to the overall energy of the star.) Whereas in Newtonian gravity theory there is now, as discussed in Sect. 1.6, a quite satisfying existence theory for this class, which can also be used to deduce some important physical properties, the situation is still embarrassingly blank in general relativity.

And the question is by no means of purely academic interest. It is a central consistency problem of general relativity and all the more so in that, according to recent studies, e.g., Babichev and Langlois (2010), alternative gravity theories like $f(R)$ theories and scalar–tensor theories have severe difficulties in providing a consistent description of strong gravity neutron stars, as they are frequently observed in the universe.

As far as we are aware, no attempts were made to clarify the existence problem of relativistic fluid stars in general relativity prior to Rendall and Schmidt (1991) and Lindblom and Masood-ul-Alam (1994), and then only for static and spherically symmetric bodies, and with specially designed, non-standard mathematical methods. In Pfister (2011), a more general proof was provided, which is based on a transcription of the Einstein equations to a system of coupled, nonlinear integral equations, and an iterative solution of this system by the standard Banach fixed point theorem. (This method was also successful for the corresponding Newtonian problem in Sect. 1.6.) This proof applies to all piecewise Lipschitz continuous equations of state $\varrho(p)$, and hereby covers at least some phase transitions, as they typically appear in strong gravity neutron stars.

However, an extension of this method to rotating bodies (and nearly all stars in the universe rotate, some of them with millisecond periods!) has not yet been achieved, although one of the best numerical analyses of such stars (Bonazzola et al. 1993), which exploits a similar mathematical scheme, gives hope for the future. Presently, the only (very partial!) existence proof for rotating stars in general relativity comes from the work of Heilig (1995). However, since this work is based on the implicit function theorem, it is so far impossible to assess how relativistic or compact these stars can be, and how fast they can rotate. For instance, this work cannot answer the astrophysically relevant question of how the Chandrasekhar limiting mass M_C (see the beginning of this section) depends on the angular momentum J and the equation of state $\varrho(p)$. [For a very recent and comprehensive overview of many facts, methods, and problems concerning rotating relativistic stars, see Friedman and Stergioulas (2013).]

In recent times, there has also been work on the existence of elastic bodies in general relativity (Park 2000) and on so-called Vlasov–Einstein systems (Rein and Rendall 1993), i.e., on a collisionless relativistic gas, which may be approximately applicable to galaxies. But again, an extension to non-spherical systems has so far only been possible (Andersson et al. 2010; Andréasson et al. 2011, 2014) in the realm of the implicit function theorem, so that the same reservation applies as to the work of Heilig above.

In the last part of this chapter, we would like to compile in a more phenomenological way some facts about gravitation (in the form of general relativity), considering situations where it differs drastically and characteristically from all other forces and phenomena in nature. The first example concerns the topic of antiparticles. Whereas all properties like electric charge, baryon number, lepton number, and strangeness are the reverse (are 'anti') for antiparticles relative to their particle partners, the mass (the 'gravitational charge') does not change sign. A first, very indirect experimental proof of this fact comes from Eötvös-type experiments, i.e., from the universality of free fall (see Sect. 1.6). Since the precision for such experiments has today reached the value 10^{-13} (Schlamminger et al. 2008), the contribution of the virtual electron–positron pairs to the mass of suitable materials would be in conflict with these experiments if the positron had negative mass. A somewhat more direct test is provided by the neutral K mesons and their interference and decay properties into two or three pions. Since the difference between the two mass eigenvalues of this system is extremely small ($\approx 10^{-10}$) in relation to the overall K mass, the system would react extremely sensitive to a negative mass of the antiparticle $\bar{K}^0$, and this possibility is largely excluded by the experiment (Good 1961). A direct test for the positive mass of antiparticles is to be expected in the near future from the anti-hydrogen atoms which can be produced at CERN, and can presumably be prepared in such a way that their free-fall properties can be measured (CERN Courier 2013).

The second example of the very special behavior of general relativity concerns the twin paradox. In special relativity, this phenomenon, already mentioned by Einstein in his foundational paper of 1905, is usually formulated in the following way: the twins start at an event e_1 and meet again at an event e_2. Whereas twin A moves directly (inertially, on a straight line in Minkowski spacetime) from e_1 to e_2, twin B executes a detour with acceleration phases, usually with a sudden turnaround at the event of greatest distance from A. Then the unequivocal result, also from the perspective of B, is that twin A is older than B at their reunion in event e_2 [see any good textbook on special relativity, or, in particular, Schild (1959)]. And this twin paradox is so important for special relativity that, with its denial, one loses all the central predictions of special relativity. What is not so often stressed is the fact that the acceleration phases of B may be caused by arbitrary forces, but not by gravitational 'forces'! Indeed, in strong gravitational fields the analogue of the twin paradox can have completely different results. The curved spacetime of general relativity has the 'singular' property (already mentioned in Sect. 1.3) that there can be more than one geodesic between the events e_1 and e_2. Even though neither twin then experiences acceleration, the aging is typically asymmetric (Holstein and Swift 1972). In the exterior Schwarzschild metric (outside the horizon at $r = 2M$), there exist even 'geodesic diangles' with arbitrary time difference between their two sides. [For a quasi-popular overview of many facts and examples in the realm of the twin paradox, and an almost complete list of references up to 1970, see Marder (1971). See also the recent paper Gasperini (2014).]

A third example of the very special behavior of general relativity comes from the interplay between gravitation, quantum mechanics, and thermodynamics, and it is one of the few examples where the (extremely small) value of the gravitational constant G plays a decisive role. It addresses the question of an analogue to the electromagnetic blackbody radiation which was, in the hands of Planck and Einstein, the midwife of quantum mechanics. As already shown in Smolin (1984, 1985), at least in linearized general relativity, gravitational waves interact so weakly with matter (fulfilling the positive energy condition) that in order to keep these waves in a box, and to reach thermal equilibrium in a finite time, so much mass must be accumulated in the walls of the box that it undergoes gravitational collapse. For all reasonable absorption mechanisms for gravitational waves (e.g., by classical matter, quantum matter, viscous matter, phonon excitation), there results for the absorption efficiency ϵ a universal formula

$$\epsilon = \left(\frac{GM}{L}\right)\left(\frac{\lambda}{L}\right)^2\left(\frac{v}{c}\right)^3 f(\vartheta) \ll 1 , \tag{3.17}$$

where M and L are the mass and the spatial extension of the 'absorber', λ is the wavelength of the gravitational radiation, v is the (sound) velocity in the absorber, and $f(\vartheta)$ is a function of the angle at which the radiation is incident on the detector. Since all factors in (3.17) are less than 1, most of them much less than 1, there does not exist any material in nature that could produce 'blackbody gravitational radiation' at a definite temperature. To quote from Smolin's papers:

> We cannot argue, as did Planck and Einstein for the case of the electromagnetic field, from the existence of an equilibrium state involving radiation and matter to the necesssity that the energy of the field is carried and transferred to matter in quanta satisfying $E = \hbar\omega$. [...] Gravitational radiation is a more degraded form of energy than is heat, as the thermal energy of a solid will be slowly converted into gravitational radiation, while in general not all of the energy of any given distribution of gravitational radiation can be converted back into heat.

In Garfinkle and Wald (1985), a quite artificial, highly charged, and nearly collapsed mass shell was proposed as a complete absorber and therefore thermal equalizer of gravitational radiation. However, in Dell (1987), it was shown that, under realistic circumstances, such a model is exposed to severe instability processes.

A last example addresses the rarely discussed fact that thermodynamics and statistical mechanics show very unusual, indeed counterintuitive behavior in systems dominated by gravity. Qualitatively, it is of course known that, e.g., the 'cosmological fluid' does not develop into a homogeneous equilibrium of fixed temperature, as is the case for all non-gravitational closed systems, but 'starting from a largely structureless initial state near the Big Bang, spontaneously' develops many types of complex structures up to the most complex structures imaginable: life on Earth. We have also seen at the beginning of this section that a pressure, usually the archetypal repulsive force, acts attractively in the interior of a heavy star, and actually contributes to its collapse. However, a more systematic study of the peculiar thermodynamic behavior of gravitation-dominated systems began relatively late, with Antonov (1962), in which it was shown for a spherical star system that the phase-space volume inside the energy shell diverges for an unshielded $1/r$ potential with sufficiently many particles.

This study was extended and deepened in Lynden-Bell and Wood (1968) and Thirring (1970), with the following strange results. The heat capacity becomes negative, with the consequence that, when two thermal systems come into contact, the hotter one loses heat and yet gets hotter, whereas the colder one gains heat and yet gets colder. Furthermore, the energy is no longer an extensive quantity in the presence of long range interactions, and gravitational systems do not possess the thermodynamic limit in the usual sense. Mathematically, the methods of the canonical and microcanonical ensembles no longer lead to equivalent results, and only the latter is consistently applicable to gravitation-dominated systems. In Lynden-Bell and Wood (1968), the instabilities caused by this strange thermodynamics in a heavy star are referred to as the 'gravo-thermal catastrophe'. For a comprehensive review (up to 1989) of the unusual thermodynamics and statistical mechanics in gravitation-dominated systems see Padmanabhan (1990).

Naturally, a consistent thermodynamics of gravitation-dominated systems necessitates a new or extended definition of the entropy expression, to which the gravitational field contributes presumably only through its Weyl tensor part, and where the total entropy (matter + gravitational field) should be an extrinsic quantity and should always be positive and monotonically increasing in time. Such a new entropy expression, for which an interesting proposal has recently been made in Clifton et al. (2013), may also contribute to Penrose's hypothesis that, in the Big Bang, the Weyl curvature vanishes (see Sect. 2.1), and to a consistent entropy definition for black holes (see above).

References

Aldersley, S.J.: Dimensional analysis in relativistic gravitational theories. Phys. Rev. D **15**, 370–377 (1977)

Andersson, L., Beig, R., Schmidt, B.G.: Rotating elastic bodies in Einstein gravity. Commun. Pure Appl. Math. **63**, 559–589 (2010)

Andréasson, H., Kunze, M., Rein, G.: Existence of axially symmetric static solutions of the Einstein–Vlasov system. Commun. Math. Phys. **308**, 23–47 (2011)

Andréasson, H., Kunze, M., Rein, G.: Rotating, stationary, axially symmetric spacetimes with collisionless matter. Commun. Math. Phys. **329**, 787–808 (2014)

Ansorg, M., Pfister, H.: A universal constraint between charge and rotation rate for degenerate black holes surrounded by matter. Class. Quantum Gravit. **25**, 035009 (2008)

Antonov, V.A.: Most probable phase distribution in spherical star systems and conditions for its existence. Vestn. Leningr. Univ., Math. Mekh., Astron. **7**, 135–146 (1962). English translation In: Goodman, J., Hut, P. (eds.) IAU Symposion 113: Dynamics of Star Clusters, pp. 525–540. Reidel, Dordrecht (1985)

Arnowitt, R., Deser, S., Misner, C.W.: The dynamics of general relativity. In: Witten, L. (ed.) Gravitation: An Introduction to Current Research, pp. 227–265. Wiley, New York (1962)

Babichev, E., Langlois, D.: Relativistic stars in $f(R)$ and scalar–tensor theories. Phys. Rev. D **81**, 124051 (2010)

Bardeen, J.M., Carter, B., Hawking, S.W.: The four laws of black hole mechanics. Commun. Math. Phys. **31**, 161–170 (1973)

Bekenstein, J.D.: Black holes and entropy. Phys. Rev. D **7**, 2333–2346 (1973)

Bizoń, P., Rostworowski, A.: On weakly turbulent stability of anti-deSitter spacetime. Phys. Rev. Lett. **107**, 031102 (2011)

Bojowald, M.: Absence of a singularity in loop quantum cosmology. Phys. Rev. Lett. **86**, 5227–5230 (2001)

Bonazzola, S., Gourgoulhon, E., Salgado, M., Marck, J.A.: Axisymmetric rotating relativistic bodies: a new numerical approach for "exact" solutions. Astron. Astrophys. **278**, 421–443 (1993)

Born, M.: Die Theorie des starren Elektrons in der Kinematik des Relativitätsprinzips. Ann. Phys. **30**, 1–56 (1909)

Bradley, M., Fodor, G., Marklund, M., Perjés, Z.: The Wahlquist metric cannot describe an isolated rotating body. Class. Quantum Gravit. **17**, 351–359 (2000)

Bray, H.L.: Proof of the Riemannian Penrose conjecture using the positive mass theorem. J. Differ. Geom. **59**, 177–267 (2001)

Brill, D., Deser, S.: Positive definiteness of gravitational field energy. Phys. Rev. Lett. **20**, 75–78 (1968)

Brill, D., Horowitz, G.T.: Negative energy in string theory. Phys. Lett. B **262**, 437–443 (1991)

Brill, D., Pfister, H.: States of negative total energy in Kaluza–Klein theory. Phys. Lett. B **228**, 359–362 (1989)

Cartan, E.: Sur les équations de la gravitation d'Einstein, J. Math. pures et appliquées **1**, 141–203 (1922)

Cartan, E.: Leçons sur la Géométrie des Espaces de Riemann, Chap. VIII. Gauthier-Villars, Paris (1928)

Carter, B.: Global structure of the Kerr family of gravitational fields. Phys. Rev. **174**, 1559–1571 (1968)

CERN Courier: ALPHA presents novel investigation of the effect of gravity on antimatter, June 2013, p. 5

Chandrasekhar, S.: The maximum mass of ideal white dwarfs. Astrophys. J. **74**, 81–82 (1931a)

Chandrasekhar, S.: Highly collapsed configurations of stellar mass. Mon. Not. R. Astron. Soc. **91**, 456–466 (1931b)

Choptuik, M.W.: Universality and scaling in gravitational collapse of a massless scalar field. Phys. Rev. Lett. **70**, 9–12 (1933)

Choquet-(Fourès-)Bruhat, Y.: Théorème d'existence pour certain systèmes d'équations aux dérivées partielles nonlinéares. Acta Math. **88**, 141–225 (1952)

Choquet-Bruhat, Y.: The Cauchy problem. In: Witten, L. (ed.) Gravitation: An Introduction to Current Research, pp. 130–168. Wiley, New York (1962)

Choquet-Bruhat, Y.: General Relativity and the Einstein Equations. Oxford University Press, Oxford (2008)

Choquet-Bruhat, Y., Geroch, R.: Global aspects of the Cauchy problem in general relativity. Commun. Math. Phys. **14**, 329–335 (1969)

Choquet-Bruhat, Y., Marsden, J.E.: Solution of the local mass problem in general relativity. Commun. Math. Phys. **51**, 283–296 (1976)

Christodoulou, D.: The formation of black holes and singularities in spherically symmetric gravitational collapse. Commun. Pure Appl. Math. **44**, 339–373 (1991)

Christodoulou, D., Klainerman, S.: The Global Nonlinear Stability of the Minkowski Space. Princeton University Press, Princeton (1993)

Chrusciel, P.T., Lopes Costa, J.: Mass, angular momentum, and charge inequalities for axisymmetric initial data. Class. Quantum Gravit. **26**, 235013 (2009)

Chruściel, P.T., Lopes Costa, J., Heusler, M.: Stationary Black Holes: Uniqueness and Beyond. Living Rev. Rel. **15**(7) (2012), and E-print arXiv: 1205.6112 [gr-qc] (2012)

Clifton, T., Ellis, G.F.R., Tavakol, R.: A gravitational entropy proposal. Class. Quantum Gravit. **30**, 125009 (2013)

Corry, L., Renn, J., Stachel, J.: Belated decision in the Hilbert–Einstein priority dispute. Science **278**, 1270–1273 (1997)

Dain, S.: Proof of the angular momentum–mass inequality for axisymmetric black holes. J. Differ. Geom. **79**, 33–67 (2008)

Dain, S.: Extreme throat initial data sets and horizon area–angular momentum inequality for axisymmetric black holes. Phys. Rev. D **82**, 104010 (2010)

Dain, S.: Geometric inequalities for black holes. Gen. Relativ. Gravit. **46**, 1715 (2014a)

Dain, S.: Inequality between size and angular momentum for bodies. Phys. Rev. Lett. **112**, 041101 (2014b)

Dain, S., Reiris, M.: Area–angular momentum inequality for axisymmetric black holes. Phys. Rev. Lett. **107**, 05110 (2011)

De Felice, A., Tsujikawa, S.: $f(R)$-Theories. Liv. Rev. Relativ. **13**, 3 (2010)

Dell, J.: On the impossibility of a box for holding gravitational radiation in thermal equilibrium. Gen. Relativ. Gravit. **19**, 171–177 (1987)

Deser, S.: Self-interaction and gauge invariance. Gen. Relativ. Gravit. **1**, 9–18 (1970)

Ehlers, J.: Über den Newtonschen Grenzwert der Einsteinschen Gravitationstheorie. In: Nitsch, J., et al. (eds.) Grundlagenprobleme der modernen Physik, pp. 65–84. Bibliographisches Institut, Mannheim (1981)

Ehlers, J., Geroch, R.: Equation of motion of small bodies in relativity. Ann. Phys. **309**, 232–236 (2004)

Ehrenfest, P.: Gleichförmige Rotation starrer Körper und Relativitätstheorie. Physik. Zs. **10**, 918–928 (1909)

Einstein, A.: Über das Relativitätsprinzip und die aus demselben gezogenen Folgerungen. Jahrbuch der Radioaktivität und Elektronik **4**, 411–462 (1907)

Einstein, A.: Über den Einfluß der Schwerkraft auf die Ausbreitung des Lichts. Ann. Phys. **35**, 898–908 (1911)

Einstein, A.: Lichtgeschwindigkeit und Statik des Gravitationsfeldes. Ann. Phys. **38**, 355–369 (1912a)

Einstein, A.: Zur Theorie des statischen Gravitationsfeldes. Ann. Phys. **38**, 443–458 (1912b)

Einstein, A.: Gibt es eine Gravitationswirkung, die der elektromagnetischen Induktionswirkung analog ist?. Vierteljahrschrift f. gerichtl. Medizin u. öffentl. Sanitätswesen **44**, 37–40 (1912c)

Einstein, A.: Zum gegenwärtigen Stande des Gravitationsproblems. Physik. Zs. **14**, 1249–1266 (1913a). English translation In: Renn, J. (ed.) The Genesis of General Relativity, vol. 3, pp. 543–568. Springer, Dordrecht (2004)

Einstein, A.: Die formale Grundlage der allgemeinen Relativitätstheorie. Sitzb. d. Preuss. Akad. d. Wiss. Math.-phys. Kl., 1030–1085 (1914)

Einstein, A.: Zur allgemeinen Relativitätstheorie. Sitzb. d. Preuss. Akad. d. Wiss. Math.-phys. Kl., 778–786, 799–801 (1915a)

Einstein, A.: Erklärung der Periheldrehung des Merkur aus der allgemeinen Relativitätstheorie. Sitzb. d. Preuss. Akad. d. Wiss. Math.-phys. Kl., 831–839 (1915b)

Einstein, A.: Die Feldgleichungen der Gravitation. Sitzb. d. Preuss. Akad. d. Wiss. Math.-phys. Kl., 844–847 (1915c)

Einstein, A.: Die Grundlagen der allgemeinen Relativitätstheorie. Ann. Phys. **49**, 769–822 (1916)

Einstein, A.: Über Gravitationswellen. Sitzb. Preuss. Akad. Wiss. Berlin, 154–167 (1918)

Einstein, A., Grossmann, M.: Entwurf einer verallgemeinerten Relativitätstheorie und einer Theorie der Gravitation. Teubner, Leipzig (1913). Reprinted, together with an important addendum in Zs. f. Math. Phys. **62**, 225–261 (1914)

Eling, C., Guedens, R., Jacobson, T.: Nonequilibrium thermodynamics of spacetime. Phys. Rev. Lett. **96**, 121301 (2006)

Feynman, R.P.: Lectures on Gravitation. Addison-Wesley, Reading (1995). Particularly lectures 3–6

Fierz, M., Pauli, W.: Relativistic wave equations for particles of arbitrary spin in an electromagnetic field. Proc. R. Soc. Lond. A **173**, 211–232 (1939)

Flanagan, E.: Hoop conjecture for black hole horizon formation. Phys. Rev. D **44**, 2409–2420 (1991)

Friedman, M.: Foundations of Space-Time Theories. Princeton Univesity Press, Princeton (1983)

Friedman, J.L., Stergioulas, N.: Rotating Relativistic Stars. Cambridge University Press, Cambridge (2013)

Friedrich, H.: Existence and structure of past asymptotically simple solutions of Einstein's field equations with positive cosmological constant. J. Geom. Phys. **3**, 101–117 (1986)

Friedrich, H.: On the AdS stability problem. Class. Quantum Gravit. **31**, 105001 (2014)

Friedrichs, K.: Eine invariante Formulierung des Newtonschen Gravitationsgesetzes und des Grenzübergangs vom Einsteinschen zum Newtonschen Gesetz. Math. Ann. **98**, 566–575 (1927)

Gabach Clément, M.E., Jaramillo, J.L.: Black hole area–angular momentum–charge inequality in dynamical non-vacuum space-times. Phys. Rev. D **86**, 064021 (2012)

Gabach Clément, M.E., Jaramillo, J.L., Reiris, M.: Proof of the area–angular momentum–charge inequality for axisymmetric black holes. Class. Quantum Gravit. **30**, 065017 (2013)

Garfinkle, D., Wald, R.M.: On the possibility of a box for holding gravitational radiation in thermal equilibrium. Gen. Relativ. Gravit. **17**, 461–473 (1985)

Gasperini, M.: The twin paradox in the presence of gravity. Mod. Phys. Lett. A **29**, 1450149 (2014)

Genzel, R., Eisenhauer, F., Gillesen, S.: The Galactic center massive black hole and nuclear cluster. Rev. Mod. Phys. **82**, 3121–3195 (2010)

Gerlach, U.H.: Derivation of the ten Einstein field equations from the semiclassical approximation to quantum geometrodynamics. Phys. Rev. **177**, 1929–1941 (1977)

Gödel, K.: An example of a new type of cosmological solution of Einstein's field equations of gravitation. Rev. Mod. Phys. **21**, 447–450 (1949)

Good, M.L.: K_2^0 and the equivalence principle. Phys. Rev. **121**, 311–313 (1961)

Greene, J.E., Ho, L.C.: Active galactic nuclei with candidate intermediate-mass black holes. Astrophys. J. **610**, 722–736 (2004)

Gundlach, C., Martín-García, J.M.: Critical phenomena in gravitational collapse. Liv. Rev. Relativ. **10**, 5 (2007)

Gupta, S.N.: Gravitation and electromagnetism. Phys. Rev. **96**, 1683–1685 (1954)

Hawking, S.W.: Gravitational radiation from colliding black holes. Phys. Rev. Lett. **26**, 1344–1346 (1971)

Hawking, S.W.: Particle creation by black holes. Commun. Math. Phys. **43**, 199–220 (1975)

Hawking, S., Ellis, G.F.R.: The Large Scale Structure of Space-time. Cambridge University Press, Cambridge (1973)

Hawking, S.W., Penrose, R.: The singularities of gravitational collapse and cosmology. Proc. R. Soc. Lond. A **314**, 529–548 (1970)

Heilig, U.: On the existence of rotating stars in general relativity. Commun. Math. Phys. **166**, 457–493 (1995)

Hennig, J., Ansorg, M., Cederbaum, C.: A universal inequality between the angular momentum and horizon area for axisymmetric and stationary black holes with surrounding matter. Class. Quantum Gravit. **25**, 162002 (2008)

Hennig, J., Cederbaum, C., Ansorg, M.: A universal inequality for axisymmetric and stationary black holes with surrounding matter in the Einstein–Maxwell theory. Commun. Math. Phys. **293**, 449–467 (2010)

Hertz, H.: Die Constitution der Materie: Eine Vorlesung über die Grundlagen der Physik aus dem Jahre 1884. Springer, Berlin (1999)

Heusler, M.: Black Hole Uniqueness Theorems. Cambridge University Press, Cambridge (1996)

Hewish, A.S., Bell, J., Pilkington, J.D.H., Scott, R.F., Collins, R.A.: Observation of a rapidly pulsating radio source. Nature **217**, 709–713 (1968)

Hilbert, D.: Die Grundlagen der Physik. (Erste Mitteilung). Nachr. d. Königl. Ges. d. Wiss. Göttingen. Math.-phys. Kl., 395–407 (1916). English translation In: Renn, J. (ed.) The Genesis of General Relativity, vol. 4, pp. 1003–1016. Springer, Dordrecht (2007)

Hilbert, D.: Die Grundlagen der Physik (Zweite Mitteilung). Nachr. d. Königl. Ges. d. Wiss. Göttingen. Math.-phys. Kl., 53–76 (1917). English translation In: Renn, J. (ed.) The Genesis of General Relativity, vol. 4, pp. 1931–1961. Springer, Dordrecht (2007)

Holstein, B.R., Swift, A.R.: The relativity twins in free fall. Am. J. Phys. **40**, 746–750 (1972)

Horowitz, G.T., Perry, M.J.: Gravitational energy cannot become negative. Phys. Rev. Lett. **48**, 371–374 (1982)

Huisken, G., Ilmanen, T.: The inverse mean curvature flow and the Riemannian Penrose inequality. J. Differ. Geom. **59**, 352–437 (2001)

Hulse, R.A., Taylor, J.H.: Discovery of a pulsar in a binary system. Astrophys. J. **195**, L51–L53 (1975)

Israel, W.: Dark stars: the evolution of an idea. In: Hawking, S.W., Israel, W. (eds.) Three Hundred Years of Gravitation. Cambridge University Press, Cambridge (1987)

Jacobson, T.: Thermodynamics of spacetime: the Einstein equation of state. Phys. Rev. Lett. **75**, 1260–1263 (1995)

Klainerman, S., Nicolò, F.: On local and global aspects of the Cauchy problem in general relativity. Class. Quantum Gravit. **16**, R73–R157 (1999)

Klein, C.: Harrison transformation of hyperelliptic solutions and charged dust disks. Phys. Rev. D **65**, 084029 (2002)

Klein, M.J., et al.: The Collected Papers of Albert Einstein, vol. 5. Princeton University Press, Princeton (1993)

Klein, M.J., et al.: The Collected Papers of Albert Einstein, vol. 4. Princeton University Press, Princeton (1995)

Kraichnan, R.H.: Special-relativistic derivation of generally covariant gravitation theory. Phys. Rev. **98**, 1118–1122 (1955)

Kramer, M., Wex, N.: The double pulsar system: a unique laboratory for gravity. Class. Quantum Gravit. **26**, 073001 (2009)

Ladyzhenskaya, O.A., Uraltseva, N.N.: Linear and Quasilinear Elliptic Equations. Academic, New York (1968)

Lifschitz, E.M., Khalatnikov, I.M.: Investigations in relativistic cosmology. Adv. Phys. **12**, 185–249 (1963)

Lindblom, L., Masood-ul-Alam, A.K.M.: On the spherical symmetry of static stellar models. Commun. Math. Phys. **162**, 123–145 (1994)

Lovelock, D.: The uniqueness of the Einstein field equations in a four-dimensional space. Arch. Ration. Mech. Anal. **33**, 54–70 (1969)

Lovelock, D.: The four-dimensionality of space and the Einstein tensor. J. Math. Phys. **13**, 874–876 (1972)

Ludvigsen, M., Vickers, J.A.G.: A simple proof of the positivity of Bondi mass. J. Phys. A **15**, L67–L70 (1982)

Lynden-Bell, D.: Galactic nuclei as collapsed old quasars. Nature **223**, 690–694 (1969)

Lynden-Bell, D., Wood, R.: The gravo-thermal catastrophe in isothermal spheres and the onset of red-giant structure for stellar systems. Mon. Not. R. Astron. Soc. **138**, 495–525 (1968)

Malament, D.: Topics in the Foundations of General Relativity and Newtonian Gravitation Theory. University of Chicago Press, Chicago (2012)

Marder, L.: Time and the Space Traveller. George Allen and Unwin, London (1971)

McConnell, N.J., et al.: Two ten-billion-solar mass black holes at the centres of giant elliptic galaxies. Nature **480**, 215–218 (2011)

Misner, C.W., Thorne, K.S., Wheeler, J.A.: Gravitation. Freeman, San Francisco (1973)

Narayan, R., McClintock, J.E.: Observational Evidence for Black Holes. E-print arXiv: 1312.6698v2 (2014)

Navarro, J., Sancho, J.B.: On the naturalness of Einstein's equation. J. Geom. Phys. **58**, 1007–1014 (2008)

Newman, E.T., et al.: Metric of rotating, charged mass. J. Math. Phys. **6**, 918–919 (1965)

Norton, J.: How Einstein found his field equations, 1912–1915. Hist. Stud. Phys. Sci. **14**, 253–316 (1984). Reprinted In: Howard, D., Stachel, J. (eds.) Einstein and the History of General Relativity. Einstein Studies, vol. 1, pp. 101–159. Birkhäuser, Boston (1989)

Norton, J.: What was Einstein's principle of equivalence? Stud. Hist. Phil. Sci. **16**, 203–246 (1985). Reprinted In: Howard, D., Stachel, J. (eds.) Einstein and the History of General Relativity. Einstein Studies, vol. 1, pp. 5–47. Birkhäuser, Boston (1989)

Ohanian, H.C.: What is the principle of equivalence? Am. J. Phys. **45**, 903–909 (1977)

O'Murchadha, N., York, J.W.: Existence and uniqueness of solutions of the Hamiltonian constraint of general relativity on compact manifolds. J. Math. Phys. **14**, 1551–1557 (1973)

Oppenheimer, J.R., Snyder, H.: On continued gravitational contraction. Phys. Rev. **56**, 455–459 (1939)

Oppenheimer, J.R., Volkoff, G.: On massive neutron cores. Phys. Rev. **55**, 374–381 (1939)

Padmanabhan, T.: Statistical mechanics of gravitating systems. Phys. Rep. **188**, 285–362 (1990)

Pais, A.: 'Subtle is the Lord …'. The Science and the Life of Albert Einstein. Oxford University Press, Oxford (1982)

Palatini, A.: Deduzione invariantina delle equazioni gravitationali dal principio di Hamilton. Rend. Circ. Mat. Palermo **43**, 203–212 (1919)

Papapetrou, A.: Spinning test particles in general relativity. I. Proc. R. Soc. Lond. A **209**, 248–258 (1951)

Park, J.: Spherically symmetric static solutions of the Einstein equations with elastic matter source. Gen. Relativ. Gravit. **32**, 235–252 (2000)

Penrose, R.: Gravitational collapse and space-time singularities. Phys. Rev. Lett. **14**, 57–59 (1965)

Penrose, R.: Gravitational collapse: the role of general relativity. Riv. Nuovo Cimento **1**, 252–276 (1969)

Penrose, R.: Naked singularities. Ann. N. Y. Acad. Sci. **224**, 125–134 (1973)

Pfister, H.: A new and quite general existence proof for static and spherically symmetric perfect fluid stars in general relativity. Class. Quantum Gravit. **28**, 075006 (2011)

Pfister, H., King, M.: The gyromagnetic factor in electrodynamics, quantum theory and general relativity. Class. Quantum Gravit. **20**, 205–213 (2003)

Polchinski, J.: String Theory, vol. II. Cambridge University Press, Cambridge (1998)

Raychaudhuri, A.: Relativistic cosmology. I. Phys. Rev. **98**, 1123–1126 (1955)

Rein, G., Rendall, A.D.: Smooth static solutions of the spherically symmetric Vlasov–Einstein system. Ann. Inst. Poincaré **59**, 383–397 (1993)

Rendall, A.D., Schmidt, B.G.: Existence and properties of spherically symmetric static fluid bodies with a given equation of state. Class. Quantum Gravit. **8**, 985–1000 (1991)

Reula, O.: Existence theorem for solutions of Witten's equation and nonnegativity of total mass. J. Math. Phys. **23**, 810–814 (1982)

Reula, O.: On existence and behaviour of asymptotically flat solutions of the stationary Einstein equations. Commun. Math. Phys. **122**, 615–624 (1989)

Rose, B.: Construction of matter models which violate the strong energy condition and may avoid the initial singularity. Class. Quantum Gravit. **3**, 975–995 (1986)

Sakharov, A.D.: Vacuum quantum fluctuations in curved space and the theory of gravitation. Sov. Phys. Doklady **12**, 1040–1041 (1968)

Schaudt, U.M.: On the Dirichlet problem for the stationary and axisymmetric Einstein equations. Commun. Math. Phys. **190**, 509–540 (1998)

Schaudt, U.M., Pfister, H.: The boundary value problem for the stationary and axisymmetric Einstein equations is generically solvable. Phys. Rev. Lett. **77**, 3284–3287 (1996)

Schild, A.: The clock paradox in relativity theory. Am. Math. Mon. **66**, 1–18 (1959)

Schlamminger, S., et al.: Test of the equivalence principle using a rotating torsion balance. Phys. Rev. Lett. **100**, 041101 (2008)

Schoen, R., Yau, S.-T.: On the proof of the positive mass conjecture in general relativity. Commun. Math. Phys. **65**, 45–76 (1979)

Schoen, R., Yau, S.-T.: Proof of the positive mass theorem. II. Commun. Math. Phys. **79**, 231–260 (1981)

Schoen, R, Yau, S.-T.: Proof that the Bondi mass is positive. Phys. Rev. Lett. **48**, 369–371 (1982)

Schulman, R., et al.: The Collected Papers of Albert Einstein, vol. 8. Princeton University Press, Princeton (1998)

Serrin, J.: The problem of Dirichlet for quasilinear elliptic differential equations with many independent variables. Philos. Trans. R. Soc. **264**, 413–496 (1969)

Shapiro, S.L., Teukolsky, S.A.: Formation of naked singularities: the violation of cosmic censorship. Phys. Rev. Lett. **66**, 994–997 (1991)

Smolin, L.: The thermodynamics of gravitational radiation. Gen. Relativ. Gravit. **16**, 205–210 (1984)

Smolin, L.: On the intrinsic entropy of the gravitational field. Gen. Relativ. Gravit. **17**, 417–437 (1985)

Sornette, D.: Critical Phenomena in Natural Sciences. Springer, Berlin (2009)

Sotirio, T.P., Faraoni, V.: $f(R)$-theories of gravity. Rev. Mod. Phys. **82**, 451–497 (2010)

Square, A., Abbott, E.E.: Flatland. A Romance of Many Dimensions. Seeley, London (1884). Reprinted, with an introduction by A. Lightman. Penguin, New York (1998)

Stephani, H., Kramer, D., MacCallum, M., Hoenselaers, C., Herlt, E.: Exact Solutions of Einstein's Field Equations. Cambridge University Press, Cambridge (2003)

Synge, J.L.: Relativity: The General Theory. North Holland, Amsterdam (1960)

Szabados, L.B.: Quasi-local energy–momentum and angular momentum in general relativity. Liv. Rev. Relativ. **7**, 4 (2004)

Taubes, C.H., Parker, T.: On Witten's proof of the positive energy theorem. Commun. Math. Phys. **84**, 223–238 (1982)

Taylor, J.H., Fowler, L.A., McCulloch, P.M.: Measurement of general relativistic effects in the binary pulsar PSR 1913+16. Nature **277**, 437–440 (1979)

Thirring, W.E.: An alternative approach to the theory of gravitation. Ann. Phys. **16**, 96–117 (1961)

Thirring, W.: Systems with negative specific heat. Zs. Physik **235**, 339–352 (1970)

Thorne, K.S.: Nonspherical gravitational collapse—a short review. In: Klauder, J. (ed.) Magic Without Magic, pp. 231–258. Freeman, San Francisco (1972)

Tolman, R.C.: Static solutions of Einstein's field equations for spheres of fluid. Phys. Rev. **55**, 364–373 (1939)

Vermeil, H.: Notiz über das mittlere Krümmungsmass einer n-fach ausgedehnten Riemann'schen Mannigfaltigkeit. Nachr. d. Ges. d. Wiss. Göttingen math.-phys. Kl., 334–344 (1917)

Wald, R.M.: General Relativity. University of Chicago Press, Chicago (1984)

Weinberg, S.: Dynamic and algebraic symmetries. In: Deser, S., et al. (eds.), 1970 Brandeis University Summer Institute in Theoretical Physics, vol. 1, pp. 283–393. MIT Press, Cambridge (1970)

Weyl, H.: Space–Time–Matter. Methuen, London (1922)

Wheeler, J.A.: Geometrodynamics. Academic, New York (1962)

Wheeler, J.A.: Gravitation as geometry. II. In: Chiu, H.Y., Hoffmann, W.F. (eds.) Gravitation and Relativity. Benjamin, New York (1964)

Wheeler, J.A.: Our universe: the known and the unknown. Am. Sci. **56**, 1–20 (1968)

Will, C.: Theory and Experiment in Gravitational Physics. Cambridge University Press, Cambridge (1993)

Witten, E.: A new proof of the positive energy theorem. Commun. Math. Phys. **80**, 381–402 (1981)

Witten, E.: Instability of the Kaluza–Klein vacuum. Nucl. Phys. **B195**, 481–492 (1982)

Witten, E.: Reflection on the fate of spacetime. Phys. Today, April 1996, 24–30

Yang, S.: On the geodesic hypothesis in general relativity. Commun. Math. Phys. **325**, 997–1062 (2014)

Chapter 4
Mach's Principle, Dragging Phenomena, and Gravitomagnetism

4.1 Early Ideas and Statements by Mach, Friedlaender, Föppl, and Einstein

Concerning E. Mach's critique of Newton's mechanics, and concerning his novel views on inertia and the relativity of rotation, reference is usually made only to Mach's famous mechanics book (Mach 1883). However, Mach had already formulated these views in the year 1868 in a seminar entitled *On some key issues of physics*, and published them as notes (pp. 47–51) in his booklet (Mach 1872), which was partly stimulated by the related booklet (Neumann 1870). Here Mach writes:

> Obviously, it is the same whether we think of the Earth as rotating around its axis, or we think of the Earth as standing still, and the celestial bodies rotating around it. Geometrically this is exactly the same case of a relative rotation between the Earth and the celestial bodies. Only the first view is astronomically easier and simpler. But if we think of the Earth at rest and the other celestial bodies revolving around it, there is no flattening of the Earth, no Foucault's experiment, and so on—at least according to our usual conception of the law of inertia. Now, one can solve the difficulty in two ways; either all motion is absolute, or our law of inertia is wrongly expressed. Neumann preferred the first supposition, I the second. The law of inertia must be so conceived that exactly the same thing results from the second supposition as from the first. By this it will be evident that, in its expression, regard must be paid to the masses of the universe. [...] Now, what share has every mass in the determination of direction and velocity in the law of inertia? No definite answer can be give to this by our experiences. We only know that the share of the nearest masses vanishes in comparison with that of the farthest. We would, then, be able completely to make out the facts known to us if, for example, we were to make the simple supposition that all bodies act in the way of determination proportionally to their masses and independently of the distance, or proportionally to the distance and so on.

In his mechanics book (Mach 1883), Mach essentially repeats the above views and hypotheses, and partly extends on them. An additional quotation from this book, which is of particular interest in connection with dragging models in Sect. 4.2, is the following:

© Springer International Publishing Switzerland 2015
H. Pfister, M. King, *Inertia and Gravitation*, Lecture Notes in Physics 897,
DOI 10.1007/978-3-319-15036-9_4

> The principles of mechanics can, presumably, be so conceived, that even for relative
> rotations, centrifugal forces arise. Newton's experiment with the rotating vessel of water
> simply informs us that the relative rotation of the water with respect to the sides of the
> vessel produces *no* noticeable centrifugal forces, but that such forces *are* produced by its
> relative rotation with respect to the masses of the Earth and other celestial bodies. No one
> is competent to say how the experiment would turn out if the sides of the vessel increased
> in thickness and mass till they were ultimately several leagues thick.

The calculations in Brill and Cohen (1966) and Pfister and Braun (1985)
precisely confirm, within general relativity, that a rotating heavy mass shell does
indeed induce in its interior the correct centrifugal (and Coriolis) forces (see also
Sect. 4.2 and Appendix B). But Mach presumably did not try to formulate, or
perhaps failed to formulate, a new version of the law of inertia consistent with
his above views. Furthermore, he did not give a recipe for eliminating Newton's
absolute space and time (as was then accomplished by Neumann and Lange, see
Sect. 1.2), and nor did he propose a new 'gravitational' force produced by moving,
e.g., rotating masses.

As far as we are aware, such an attempt was first formulated by the Friedlaender
brothers in Friedlaender and Friedlaender (1896). This booklet was decisively
influenced by Mach's mechanics book (even though I. Friedlaender says at one
point "Without knowing that this had already been done by Mach"), but it also
presents ideas and views going far beyond Mach's work, and even partly anticipating
ideas developed much later by Einstein. In accordance with Mach, the Friedlaenders
demand an improved form of the law of inertia such that the centrifugal force is
explicable through relative motions alone, without resorting to absolute motion. But
going beyond Mach, they propose a solution of this problem in connection with a
generalization of the law of gravitation. In the words of I. Friedlaender:

> It seems to me that the correct form of the law of inertia will only have been found when
> *relative inertia* as an effect of masses on each other and *gravitation*, which is also an effect
> of masses on each other, have been derived on the basis of a *unified law*.

And there is a footnote:

> In this connection it is greatly to be desired that the question of whether Weber's law
> [Weber's action-at-a-distance law of 1846 combined Coulomb's law with Ampère's law
> for electric currents, and was later superseded by Maxwell's field theory] is to be applied
> to gravitation and also the question of the propagation velocity of gravitation should be
> resolved.

At the end of the booklet B. Friedlaender even vaguely anticipates the incorpo-
ration of inertia and gravity into the properties of space and time (i.e., Einstein's
principle of equivalence) by saying:

> It is also readily seen that in accordance with our conception the motions of the bodies
> of the Solar System can be regarded as pure inertial motions, whereas in accordance with
> the usual conception the inertial motion, or rather its gravitationally continually modified
> tendency strives to produce a rectilinear tangential motion.

Quite remarkably, I. Friedlaender also conceived and executed the first experiment for the dragging of inertial frames by rotating masses: a heavy, rapidly rotating
fly-wheel with a torsion balance in line with its axis. Although a dragging of the

torsion balance should be present here in principle, the effect is very many orders of magnitude below measurability, even with modern techniques. [The remarkable and innovative booklet (Friedlaender and Friedlaender 1896) seems to be the only lasting contribution of these authors to the foundations of physics. Immanuel Friedlaender, although he had a Ph.D. in physics, later became a famous vulcanologist, founding and directing the world's first institute for vulcanology in Naples, with a donation to the ETH Zurich, existing until today. Benedict Friedlaender was a zoologist and expert in the science of sexual reproduction.]

An independent and in principle more promising and more professional dragging experiment was described in Föppl (1904a): the rotation of the whole Earth should influence the axis of a gyroscope consisting of two heavy fly-wheels, a primitive forerunner of the Stanford Gravity Probe B experiment (see Sect. 4.4). Unfortunately, this experiment had only an accuracy of 2 % of the angular velocity of the Earth, whereas we know today (see Sect. 4.4) that an accuracy of at least 10^{-9} of this angular velocity would be necessary for a positive effect. Mach reacted positively to these experiments by Friedlaender and Föppl in later editions of his mechanics book. In a related theoretical paper (Föppl 1904b), Föppl expresses the opinion that the connection between the local inertial systems and the fixed stars cannot be accidental. He focused on the Coriolis forces exerted by rotating masses (in contrast to all other authors, who considered only the centrifugal forces), but surprisingly he denied that these 'velocity forces' were a type of gravitational force.

The first paper by Einstein which explicitly addresses Machian questions, and which also says that Mach's mechanics book was a decisive motivation for some of his own attempts, is obviously Einstein (1912). Although this paper contains many new and interesting ideas and models (e.g., a spherical mass shell which is still useful today in general relativity as a substitute for the Newtonian mass point, the first calculation of a dragging effect in a relativistic gravity theory, and the hypothesis of a mass increase due to surrounding heavy masses), most of the details of this paper are (in retrospect) wrong. This begins with the title *Is there a gravitational action analogous to electromagnetic induction?* Since the paper Einstein (1912) is based on a scalar relativistic gravity theory, it can never produce an action similar to the one resulting from the vectorial structure of electrodynamics. Furthermore, even in the final version of general relativity, there exists (in the linear approximation) a gravitomagnetic action analogous to Ampère's law for electric currents, but no action analogous to Faraday's law of induction (see Sect. 4.4).

All the details of Einstein (1912) are based on Einstein's 'derivation' of the mass increase of a test mass m due to the presence of a nearby heavy mass M, e.g., in the form of a mass shell. However, in general relativity it turned out, after numerous controversial claims, that such a mass increase is only an untestable coordinate effect (Brans 1962). Finally, in general relativity there does indeed exist a dragging of test masses inside a linearly accelerated mass shell (mass M, radius R), but with a dragging factor $4M/3R$ (Pfister et al. 2005), in contrast to Einstein's result $3M/2R$.

In June 1913, Einstein wrote a famous letter to Mach (see, e.g., Misner et al. 1973, pp. 544–545) in which he says: "It necessarily follows [in the so-called Entwurf theory] that *inertia* has its origin in a type of *interaction* of the bodies,

just as in your consideration concerning Newton's experiment with the bucket."
In Einstein (1913), again based on the Entwurf theory and on the Einstein–Besso
manuscript from June 1913 (Klein et al. 1995, pp. 344–473), Einstein particularly
stresses the idea that the relativity of inertia is a natural result of his theory. In this
tensorial theory he is now also able to calculate the dragging of test masses inside a
rotating mass shell, due to a Coriolis force, by analogy with the vector potential of
electric currents. Only the magnitude of this effect is smaller by a factor 1/2 than in
the final version of general relativity.

The expression 'Mach's principle' appears first in the paper Einstein (1918) with
the words "The G-field [gravitational field] is *completely* determined by the masses
of the bodies", and "according to the gravitational field equations no G-field is
possible without matter". However, Einstein then observes that, even for $T_{\mu\nu} \equiv 0$,
a non-trivial G-field $g_{\mu\nu}$ = const. is possible for all μ, ν, e.g., the Minkowski
metric. He tries to remedy this 'failure' by adding a cosmological term $\Lambda g_{\mu\nu}$ to
the field equations (Einstein 1917). But shortly after this, deSitter proved (de Sitter
1917) that in this extended theory there are also non-trivial solutions for $T_{\mu\nu} \equiv 0$.
A complete determination of the gravitational field from a given distribution of
energy–momentum $T_{\mu\nu}$ is possible, if at all, only by supplementing a given $T_{\mu\nu}$
by appropriate, e.g., cosmological, boundary conditions. (Compare the remarks
concerning the solution manifold of the gravitational field equations in Sect. 3.1.)
A further, more fundamental critique on Einstein's above formulation of Mach's
principle comes from the following fact: in order for the specification of a tensor
$T_{\mu\nu}$ to make any sense, a metric $g_{\mu\nu}$ must already be given, i.e., the statement that
the matter by itself determines the metric is meaningless.

Somewhat later, in the Princeton lectures from May 1921, published in Einstein
(1922), Einstein comes back to such Machian questions in general relativity, and
here he is more explicit and more realistic. He says:

What has to be expected along the line of Mach's thought?

1. The inertia of a body must increase when ponderable masses are piled up in its
   neighborhood.
2. A body must experience an accelerating force when neighboring masses are accelerated,
   and, in fact, the force must be in the same direction as the acceleration.
3. A rotating, hollow body must generate inside of itself a Coriolis field, which deflects
   moving bodies in the sense of the rotation, and a radial centrifugal field as well.

We will now show that these three effects, to be expected according to Mach's thoughts,
really have to be present in our theory [general relativity], although in so small magnitude
that a confirmation by laboratory experiments cannot be thought of.

We have already pointed out above that the first demand does not make sense,
but that the second and third demands are realized in general relativity. Somewhat
later in Einstein (1922), Einstein formulates concerning the above three points the
following: "We must see in them a strong support for Mach's ideas as to the relativity
of all inertial actions", but again he exaggerates in an illicit way:

If we think these ideas through consistently to their logical conclusion, we must expect
the *whole* inertia, i.e., the *whole* $g_{\mu\nu}$ field to be determined by the matter of the universe,

and not mainly by the boundary conditions at infinity. [...] Consistent with the Machian thought is only a spatially closed (finite) world, and not a quasi-Euclidean infinite one. Anyhow, it is intellectually more satisfying if the mechanical and metrical properties of space are completely determined by the matter, something that is only realized in the case of a spatially closed world.

In later years Einstein moved further and further away from the Mach principle, and finally came to a complete repudiation of it, when he wrote in a letter to F. Pirani in February 1954 (Pais 1982, p. 288): "As a matter of fact, one should no longer speak of Mach's principle at all."

Here we shall take a more positive, but nevertheless realistic and practical viewpoint concerning this principle. In contrast to, e.g., J. Barbour, we do not see "general relativity as a perfectly Machian theory" (Barbour and Pfister 1995, pp. 214–231). We see this principle rather as applying mainly to our actual universe, or a reduced class of cosmological solutions (whether spatially closed or open) of Einstein's field equations, which agree with all observations. [Hereby we disregard cosmological models like the Gödel solution (Gödel 1949) and the Ozsváth–Schücking solution (Ozsváth and Schücking 1962).] And clearly Mach had only such models in mind when he formulated his thoughts, and only Machian or anti-Machian properties of such models have any chance of being tested experimentally. In the next section, we will see that different types of models for the dragging of inertial systems do indeed realize some of Mach's ideas, and in Sect. 4.3 we investigate how far the relation between local inertial systems and the frame defined by the distant cosmic masses and the cosmic background radiation actually satisfies Machian claims.

4.2 Dragging Phenomena in General Relativity

Historically, as shown in the last section, Einstein introduced first the model of a spherical, thin mass shell as a substitute for all the matter in the universe to calculate what are nowadays known as 'dragging effects' in the framework of a scalar relativistic graviational theory and also the Entwurf theory, both of which preceded the final version of general relativity. These effects refer to the influence of accelerated masses on the local inertial frames, and in this respect they exhibit 'Machian' properties of Einstein's theory of gravitation. Within these shell-type models of 'rotating skies', one tries to reveal the true nature and origin of inertia, and in particular to answer questions about the induction of Newtonian 'fictitious' forces, such as the Coriolis and centrifugal forces. Therefrom, it can be made plausible that the motions of bodies, and especially rotation, should have only relative meaning in physics, and that it should be impossible to decide, in principle, whether an observer is rotating relative to the fixed stars, or all the distant stars and galaxies in the universe are rotating relative to the observer. It is mainly this aspect linked with the inertial structure of spacetime which we shall concentrate on in connection with the whole, complex, and in part controversial subject of

'Mach's principle' [see Barbour and Pfister (1995) for a comprehensive survey], and the extensive field of dragging phenomena in general relativity. We hope to give a new synopsis of this theme, with this specific focus not entering most textbook presentations on the foundations of gravitation and general relativity—and in a way amend Ciufolini and Wheeler's view on gravitation and inertia (Ciufolini and Wheeler 1995).

In this section we follow the route from the classic papers of Thirring and Lense of 1918 on these dragging phenomena (the so-called Lense–Thirring effect, valid in the weak-field regime of gravity) to more elaborate generalizations to strong gravitational fields and higher orders in the angular velocity ω of the rotation of the shell (focusing especially on the so-called centrifugal force problem). However, it has to be stressed that such dragging phenomena are not restricted to (mainly stationary) rotational accelerations (i.e., effectively time-independent systems), but also show up in the case of linearly accelerated bodies (i.e., in true dynamical situations), resulting in the hypothesis of a 'quasi-global equivalence principle in general relativity' between general acceleration fields and gravitational fields. We add a few remarks on dragging properties of more realistic relativistic configurations, namely rigidly rotating discs and equilibrium stellar models.

We conclude this section with more recent issues on the 'electromagnetic Thirring problem': as stated above, the standard Thirring problem describes the (nonlocal) influence of rotating masses on the inertial properties of spacetime, especially the so-called dragging of inertial frames inside a rotating mass shell relative to the asymptotic frames. It is then natural to ask whether and how properties other than inertial ones are also influenced by rotating masses, and the first (noninertial and nongravitational) properties which come to mind here are surely electromagnetic phenomena. Since the coupled Einstein–Maxwell equations are structurally not much more complicated than the pure Einstein equations (see Sect. 4.4 in the linearized case), an extension of the Thirring problem to electromagnetic phenomena is technically manageable. For instance, it can be shown that a rotating mass shell induces (to first order in ω) to a charge in its interior a dipolar magnetic field, a further 'Machian' aspect of the relativity of rotation.

Einstein's formulation of Mach's principle as formulated in the Princeton lectures of May 1921 (especially items 2 and 3, listed in the last section) established a concrete program regarding the induction of 'fictious forces' in a relativistic theory of gravity. The first attempt to calculate any such dragging effect within the final version of general relativity, effects which had already been calculated by Einstein (and Besso) in 1912–13 in the Entwurf theory (see the previous section), was made by H. Thirring (and J. Lense), starting in April 1917 with an extensive notebook entitled *Effects of rotating masses* (Thirring 1917) and finally culminating in the two well-known papers of 1918 (Thirring 1918; Lense and Thirring 1918). For Einstein's 1913 model of an infinitely thin, spherical shell with mass M, radius R, and angular velocity ω, Thirring derived to first orders in M/R [i.e., in the weak-field approximation of Einstein's equations of 1916 (Einstein 1916)] and ω a Coriolis-type force with 'dragging factor' $d_1 = 4M/3R$. To second order in ω, an additional force showed up which was treated by Thirring as a centrifugal force, although it also had an axial component and could not be made zero in the same rotating frame in which the Coriolis-type force vanishes.

In detail, calculation of the geodesic equation for test masses with (small) mass m, and with (small) velocity $\mathbf{v}$ inside and near the center ($r \ll R$) of the stationary and slowly rotating mass shell exhibits a dragging acceleration field

$$\mathbf{a} = -2d_1(\boldsymbol{\omega} \times \mathbf{v}) - d_2\big[\boldsymbol{\omega} \times (\boldsymbol{\omega} \times \mathbf{r}) + 2(\boldsymbol{\omega} \cdot \mathbf{r})\boldsymbol{\omega}\big] , \tag{4.1}$$

with $d_1 = 4M/3R$ and $d_2 = 4M/15R$ (after the correction by Laue and Pauli in 1920, see Thirring 1921). Comparing (4.1) with (1.14) of Sect. 1.6 in the stationary case (i.e., $\dot{\omega} = 0$ and no Euler forces) shows that the first term in (4.1) leads (except for the factor d_1, which is not generally equal to 1) to the Coriolis force of Newton's theory in a rotationally accelerated frame, viz.,

$$F_{\text{cor}} = -\frac{8mM}{3R}\boldsymbol{\omega} \times \mathbf{v} ,$$

and the second one to a kind of 'centrifugal force'

$$F_{\text{centr}} = -\frac{4mM}{15R}\big[\boldsymbol{\omega} \times (\boldsymbol{\omega} \times \mathbf{r}) + 2(\boldsymbol{\omega} \cdot \mathbf{r})\boldsymbol{\omega}\big] ,$$

with an axial component $\sim 2(\boldsymbol{\omega} \cdot \mathbf{r})\boldsymbol{\omega}$ that does not occur in its Newtonian counterpart and which also could not be explained in a satisfactory manner by Thirring. (For an elimination of this erroneous additional component in the centrifugal force, see below.)

But Thirring's centrifugal force term suffers from another inconsistency: as has been shown in Lanczos (1923), Thirring's model of a shell made of dust violates Einstein's field equations, because, disregarding any stresses in the shell material, the (pressureless) energy–momentum tensor is obviously not divergence free. Therefore, it does not satisfy the local energy–momentum conservation law $T^{\mu\nu}_{;\nu} = 0$ to second order in ω. (Physically, in order for the mass elements of the shell to be able to rotate on spherical orbits, the centrifugal forces have to be compensated for by appropriate stresses in the shell material.) More than three decades later, Bass and Pirani (1955) partly repeated Lanczos' arguments, but presented them in more mathematical detail, and generalized Thirring's model to a latitude-dependent mass density of the shell. [At the same time, and obviously independently, Hönl and Maue (1956) derived similar but less complete results.]

Another argument, calling for a treatment of the rotating mass shell at least up to order M^2 (or even exact in M in order to make Newton's vessel and the mass shell "several miles thick", and therefore to account for a substantial part of the whole universe), was presented in Soergel-Fabricius (1961). As already discussed in Thirring (1918), it is possible to eliminate the Coriolis acceleration inside the rotating mass shell by a transformation to an appropriately rotating reference system, i.e., a frame rotating with an opposite angular velocity $\tilde{\omega} =$

$(-4M/3R)\omega$. However, the centrifugal acceleration $(\sim M^2\omega^2/R^2)$ can vanish in the same reference system, as it should according to Mach's demand for a relativity of rotation, at best if it is of order $(M\omega/R)^2r$, rather than of order $(M\omega/R)\omega r$ in Thirring (1918).

For the exterior gravitational field of a slowly rotating spherical body, e.g., the Earth or the Sun, Thirring and the mathematician Lense (Lense and Thirring 1918) calculated the Coriolis acceleration $\mathbf{a}_{cor}$ for a test mass of velocity $\mathbf{v}$ for values $r/R \gg 1$, i.e., in the far field of the rotating source:

$$\mathbf{a}_{cor} = 2\mathbf{v} \times \mathbf{H} \,, \quad \text{with} \quad \mathbf{H}(\mathbf{r}) = \frac{2MR^2}{5r^3}\left[\boldsymbol{\omega} - \frac{3\mathbf{r}}{r^2}(\boldsymbol{\omega}\cdot\mathbf{r})\right], \tag{4.2}$$

where $\mathbf{H}$ is what is now called the 'gravitomagnetic' dipole field (for the calculation and experimental verification of $\mathbf{H}$, see Sect. 4.4). More elegantly, it is already clear from symmetry considerations that a first-order rotational perturbation of a spherical system can only produce a pure dipole field proportional to r^{-3} (see also the electromagnetic case below).

It is less well known that the Lense–Thirring papers owe nearly all their physically interesting results and the correct calculations to a correspondence with Einstein (Schulmann et al. 1998, Docs. 361, 369, 401, and 405) and Einstein's talk at the Vienna congress in 1913 (Einstein 1913), which Thirring attended. Einstein's decisive impact and contribution on the genesis of these results has been revealed extensively in Pfister (2007, 2010). In order to be historically correct and fair concerning the respective merits of Einstein, Thirring, and Lense in the discovery of (4.2), which is now known as the Lense–Thirring effect, the conclusion is that it should be called the Einstein–Thirring–Lense effect.

If one tries to reveal any influence of the cosmos on local physics in a Machian view, one surely has to consider a cosmological setting, i.e., in particular to go beyond weak gravitational fields, which are the basis for the above-mentioned calculations. Further development of the work of Einstein and Thirring had to wait almost 50 years until, in 1966, Brill and Cohen (1966) succeeded in extending Thirring's calculations to arbitrary values of M/R (but still only to first order in the angular velocity of the shell) by considering a rotational perturbation, not of Minkowski spacetime, but of the Schwarzschild solution. They derived (for the whole flat interior of the shell) a Coriolis-type acceleration, with dragging factor [compare with (4.1) and (B.17) of Appendix B]

$$d_1 = \frac{4M(4R - M)}{(2R + M)(6R - M)}, \tag{4.3}$$

where R denotes the shell radius in isotropic coordinates. (In Schwarzschild coordinates the expression would be somewhat more involved.) In the weak-field limit $M/R \ll 1$, this dragging factor coincides of course with Thirring's result $d_1 = 4M/3R$. Their central new result was that, in the collapse limit $M/2R \to 1$ of the mass shell, the dragging factor d_1 in (4.3) attains the value 1. Inertial systems inside the mass shell are dragged along with the *full* angular velocity ω of the shell. In this limit, geometrically, the interior of the mass shell is cut off from the exterior of the spacetime as a type of a separate 'universe', and one gets total dragging for interior test particles and local inertial frames.

A spherical mass shell in the collapse limit is admittedly a rather simplified cosmological model, and one may object that in this limit the shell material has somewhat unphysical properties, e.g., the stresses diverge. The usual energy conditions are already violated before the collapse, e.g., for $R < 3M/4$, the dominant energy condition fails. Nevertheless, such a configuration may be regarded as a not too unrealistic substitute for the cosmic mass distribution compared, e.g., with cylindrical mass shells. (For an extension to more realistic cosmological spacetimes, see the next section.) This result confirms (at least partly) that for physically reasonable models within general relativity, the Machian postulate of the relativity of rotation is satisfied (to first order in ω), so the classic work of Brill and Cohen may be judged as the most important positive contribution to Mach's question so far.

An extension of the results of Brill and Cohen to higher orders in ω, and in particular the long-standing problem of the induction of a correct centrifugal force by rotating masses had to wait for another 19 years to be solved in Pfister and Braun (1985) (these calculations with some minor corrections are repeated in Appendix B). The solution is based on two 'new' observations which could and should have been made already in Thirring's time, but which, for inexplicable reasons, were overlooked by all authors before 1985 (see also Pfister 2007):

- Any physically realistic, rotating body will suffer a centrifugal deformation to order ω^2 and higher, and cannot be expected to keep its spherical shape.
- If we aim and expect to realize quasi-Newtonian conditions with the 'correct' Coriolis and centrifugal forces (and no other forces!) in the interior of the rotating mass shell, this interior obviously has to be a flat piece of spacetime. To first order in ω, this flatness is more or less trivial because the only non-Minkowskian metric component $g_{t\varphi}$ is constant there, i.e., we have a constantly rotating Minkowski metric, and therefore a structurally correct Coriolis force. In contrast, to order ω^2, this flatness is by no means trivial, and it is indeed violated for Thirring's solution due to the axial component of his 'centrifugal force'. Moreover, if Thirring had extended his calculations to orders $\omega^3, \omega^4, \ldots$, he would have obtained additional forces in the interior of the rotating mass shell, in conflict with Newtonian physics in a (stationary) rotating reference system [see (1.14) of Sect. 1.6].

With these observations, the problem of the correct centrifugal force inside a rotating mass shell boils down to the question of whether it is possible to connect a 'rotating' flat interior metric through a mass shell (with, to begin with, unknown geometrical and material properties) to the non-flat but asymptotically flat exterior metric of a rotating body. In full generality, this would represent a mathematically quite intricate free boundary value problem for the stationary and axisymmetric Einstein equations. However, if one confines oneself to a perturbation expansion in the angular velocity ω, all metric functions can be expanded in spherical harmonics, i.e., due to the axial symmetry, just in Legendre polynomials $P_l(\cos\theta)$, where to order ω^n the index l is limited by $l \leq n$. In this way, the Einstein equations reduce to a system of ordinary linear differential equations for the functions $f_l^{(i)}(r)$ multiplying $P_l(\cos\theta)$ ($i = 1,\ldots,4$, for the four different metric coefficients describing the stationary, axisymmetric spacetime in the exterior of the mass shell).

In summary, by allowing for a nonspherical form of the rotating mass shell, for a nonspherical (latitude-dependent) mass distribution on it, and for differential rotation, it was shown in Pfister and Braun (1985, 1986) that, for given parameters M, R, and $\omega R \ll 1$, there exists exactly one quasi-spherical rotating mass shell (as a unique solution to the above-mentioned stationary and axisymmetric system of Einstein's equations) which induces flat geometry in its whole interior to all orders ω^n, and therefore correct Coriolis and centrifugal forces with no additional spurious forces. In short, there exists a finite region of spacetime with the 'correct' inertial structure known from Newtonian physics. Only in the collapse limit is the rotating shell with flat interior spherical and rigidly rotating, and it produces the Kerr geometry in the exterior, as was already deduced in De La Cruz and Israel (1968). For a mass shell which deviates from sphericity, even to zeroth order in ω, there is no solution with flat interior (Pfister 1989). From this one may conclude with Pfister and Braun (1985) regarding the long-standing centrifugal force problem:

> In this way, Mach's ideas on the relativity of rotation (not the whole so-called Mach principle, as stated by Einstein (1918)!) are materialised in general relativity as completely as one could ever hope within the model of a shell-type sky.

Historically, the first dragging effect, investigated within a preliminary scalar relativistic graviational theory, was of translational origin: for a test mass inside a linearly accelerated mass shell with acceleration Γ, mass M, and radius R, Einstein deduced two effects in 1912 (Einstein 1912). First, an increase in the inertial mass by a factor $1 + GM/Rc^2$ (see item 1 in Sect. 4.1), and second, a linear dragging of test bodies inside the shell with the ratio $d_{\text{linear}} = \gamma/\Gamma = 3GM/2Rc^2$. [The dragging by a linearly accelerated mass shell is calculated in the framework of the Entwurf theory in the Einstein–Besso manuscript (June 1913) on the motion of the perihelion of Mercury (Klein et al. 1995, pp. 436–437).] It is then natural to ask whether and how all the dragging effects in general relativity derived for rotating mass shells carry over from a rotational acceleration to a linear acceleration, as initiated by Einstein.

A quite general and severe problem with linearly accelerated bodies is that they need, in contrast to rotating bodies, a perpetual supply of energy in order to maintain the acceleration. And since in general relativity the equations of motion of bodies are already contained in the field equations (see Sect. 3.2), the energy source (or the 'motor' of the accelerated system) has to be included in the considered system in order to obtain a self-consistent problem. This difficulty may be the reason why, besides the historical Einstein paper (Einstein 1912), only a few articles (Farhoosh and Zimmermann 1980; Gron and Eriksen 1980; Lynden-Bell et al. 1999) have yet treated (or claimed to treat) dragging effects due to linearly accelerated masses. And even these papers compare only rather poorly with the rotating systems considered in Thirring (1918), Brill and Cohen (1966), Pfister and Braun (1985) because they treat only the weak-field case or contain special relations between mass M and charge q of the shell. Furthermore, in some of these papers the source of acceleration is not really fixed, or is removed to infinity, with the consequence that the equations of motion are in danger of being violated. And in none of these models is it guaranteed that the geometry inside the shell is flat, so that the putative dragging effects cannot be clearly distinguished from local gravitational effects due to curvature.

To overcome all these difficulties and pitfalls, a spherical Reissner–Nordström shell of nearly arbitrary mass M, charge q, and radius R was considered in Pfister et al. (2005), and a (first-order) translational acceleration of this shell was calculated, the source of this linear acceleration being a (weak) momentarily static, dipolar charge distribution $\sigma(r)$ outside the shell. The coupled Einstein–Maxwell equations in the electro-vacuum regions have a flat solution inside the shell, and for an appropriate asymptotic fall-off behavior of $\sigma(r)$, the solution is also asymptotically flat (as in the rotational examples considered so far). Three main results could be inferred from this model system. First, within the weak-field regime $M/R \ll 1$, and $q/R \ll 1$, one finds that, for the simplest power law charge distribution $\sigma(r) \sim r^{-5}$ having a finite dipole moment, the dragging factor inside the shell, calculated from the geodesic equation for neutral test particles in this region, coincides (by accident) with Thirring's value $4M/3R$. Second, for a shell with arbitrary mass but small charge, one finds that, for $\sigma(r) \sim r^{-5}$, the dragging factor d_{linear} has a similar dependence on M/R as for the rotating mass shell in Brill and Cohen (1966). Third, in the important collapse limit $2M/R \to 1$, one once again obtains $d_{linear} \to 1$, i.e., total dragging, and this for arbitrary charge distributions $\sigma(r)$.

In summary, it seems to be evident that, for 'small' (first-order) linear and rotational accelerations of a mass shell, the interior of this shell can be kept flat, and that the dragging effects in this shell exactly mimic the corresponding well-known 'inertial forces' in accelerated reference systems in Newtonian physics. Since general accelerations can (in principle) be combined from appropriate linear and circular accelerations, this gives very good arguments for the validity of a *quasi-global equivalence principle* in general relativity, a hypothesis first formulated in Pfister and Braun (1985):

> If some finite laboratory (a flat region in spacetime) is in arbitrary (weak) accelerated motion relative to the fixed stars, then all motions of free particles and all physical laws, measured from laboratory axes, are modified by inertial forces. It is argued that exactly the same modified motions and laws can be induced (at least for some time) at all places in a laboratory at rest relative to the fixed stars, by suitable and suitably moving masses outside the laboratory (e.g., in a mass shell).

In short, this hypothesis may be phrased as follows (Pfister 2014): "Every acceleration field can be understood as a gravitational field." In this connection it may be remarked that, even at the dawn of general relativity, in the years 1912–1913, similar ideas arose in discussions between Einstein and Ehrenfest (Klein et al. 1993, Docs. 409 and 411) and between Einstein and Mie (Einstein 1913). But at that time the participants were quite sceptical about such a 'macro-equivalence'.

The delta-type shell structure of the models looked at so far may be criticized as being rather unrealistic, in particular compared with isolated bodies in nature like stars and galaxies. However, there are also quite general results for the dragging behavior of isolated equilibrium stellar models with differential rotation, and rigidly rotating disk solutions. In general, for stationary axisymmetric and asymptotically flat solutions of Einstein's equations (with two Killing vectors $\xi^\mu = \partial_t$ and $\eta^\mu = \partial_\phi$), the dragging of inertial frames, as seen by asymptotic observers, is prescribed by a time-independent gravitational potential $A(\rho, z) = -\xi^\mu \eta_\mu / \eta^\mu \eta_\mu$. Continuing along the lines of earlier work in Hansen and Winicour (1975, 1977) and in Lindblom (1978) on rotating star configurations, it was shown in Pareja (2004a,b) that, if the distribution of the angular velocity of the fluid is non-negative, i.e., $\omega \geq 0$ (and nontrivially, $\omega \not\equiv 0$), then the dragging rate is positive everywhere, i.e., $A > 0$, so it has the same sign. On the other hand, for quite general differentially rotating stellar models satisfying the weak energy condition, the dragging rate is always less than the fluid's angular velocity, and hence the star has a positive angular momentum.

Rigidly rotating dust disks are degenerate limiting cases of fluid bodies of vanishing pressure and serve as more or less realistic models for certain galaxies and accretion disks. Such disk-shaped matter distributions are one of the few examples of explicit solutions of Einstein's equations being exact in the angular velocity of the rotation (like the Kerr–Newman black hole solutions mentioned in Sect. 3.3). We refer here to the dust disk of Neugebauer and Meinel (1994) and the stationary counterrotating dust disks of Klein (2001), the latter solution being in a sense a generalization of the Poisson integral to the relativistic case. Concerning their detailed dragging behavior (Meinel and Kleinwächter 1995; Frauendiener and Klein 2001) both disks, like the mass shells mentioned above in the collapse limit, show complete dragging (viewed from the asymptotic regime) in the 'ultrarelativistic limit', i.e., in the limit where the redshift diverges.

However, dragging phenomena, these non-Newtonian predictions of general relativity considered so far, are not restricted strictly to gravity. Mach's demand on the relativity of rotation exhibits some further, electromagnetic aspects, viz., the so-called electromagnetic Thirring problems, a phrase coined by Ehlers and Rindler in the papers Ehlers and Rindler (1970, 1971). Such problems try to answer the question as to whether and how properties other than inertial ones, and in particular electromagnetic properties, are also influenced by rotating masses. This is all the more important in that electromagnetism is, besides gravity, the second fundamental classical interaction in nature. In the words of Rindler (1969, p. 16):

> By its denial of AS [absolute space], Mach's principle actually implies that not only gravity but *all* physics should be formulated without reference to preferred inertial frames. It advocates nothing less than the total relativity of physics. As a result, it even implies interactions between inertia and electromagnetism. Consider, for example, a positively charged, nonconducting sphere which rotates. Each charge on it gives rise to a circular current and thus to a magnetic field. [...] By analogy again, a minute magnetic field should arise within any massive rotating shell with stationary charges inside it.

This conjecture of a gravitationally induced magnetic (dipole) field was first considered by Hofmann (1962) and later on by Cohen (1966). The influence of a rotating mass shell on electromagnetic phenomena, especially on charges sitting inside the mass shell, is usually considered within a general class of two-shell systems (see Ehlers and Rindler 1970, 1971; King and Pfister 2001): a charged shell with radius $a > 0$ (in order to avoid singularities of the electrostatic energy of the Coulomb field due to point charges) within a mass shell with radius $R > a$ is discussed in different approximations of mass M, charge q, and usually to first order in the angular velocities ω_a and ω_R of the two (inner and outer) concentric shells, resulting in stationary and axisymmetric charge and mass distributions (and therefore in time-independent gravitational and electromagnetic fields of the coupled Einstein–Maxwell equations). Mathematically, this problem amounts to first-order rotational (dipole) perturbations of three matched Reissner–Nordström metrics in the inner, intermediate, and outer regions of the two-shell system.

Now, Hofmann (1962) considered a charged shell within a rotating mass shell with radius R to first order in the mass M, charge q, and angular velocity ω, and obtained, as one would expect on Machian grounds, a magnetic dipole field induced by the rotating mass shell. For $r < R$, this field is constant along the axis of rotation, and for $r > R$ it falls off asymptotically as r^{-3}. Later on, Cohen (1966) considered a similar system exactly in M and to first order in q (i.e., there is no back-reaction of the charges on the spacetime geometry), but now with an angular velocity for the inner, charged shell. As a result of the electromagnetic test field approximation used here, the whole space inside the mass shell always stays flat (as in the pure gravitational Thirring problem), and in this way allows for a Mach-equivalent situation in non-relativistic physics. Only in this special case can one expect, if at all, to see close analogies with results from classical electrodynamics. Mathematically, this has the simplifying consequence that gravitational and electromagnetic effects decouple, and one has to solve only the Maxwell equations on a rotationally disturbed Schwarzschild background to first order in the angular velocity ω. As the mass shell approaches its collapse limit, and can again be considered as an idealized substitute for the overall masses in our universe, Cohen gets the completely Machian result that:

> [...] one cannot distinguish (even with electromagnetic fields reaching beyond the mass shell) whether the charged shell is rotating or the mass shell is rotating in the opposite direction.

The constant magnetic field inside the charged shell has the following component in the direction of the axis of rotation (in isotropic Reissner–Nordström coordinates):

$$B_z = \frac{8q}{3a} (\overline{\omega}_a - \overline{\omega}_R) \left[1 - \left(\frac{a}{R} \right)^3 \right]. \qquad (4.4)$$

Here, an overbar indicates the time dilatation between an inner and an outer observer, i.e., the relative angular velocity $\overline{\omega}_a - \overline{\omega}_R$ between the bulk of the matter of the universe and the charged shell is measured with respect to the interior proper time. As a consequence, a finite angular velocity, as seen from an observer inside the mass shell, appears to be infinitely slowed down by an asymptotic observer outside the shell. As discussed in King and Pfister (2001) within a more comprehensive class of charged two-shell systems comprising all the model systems and their approximations considered hitherto as special cases, the magnetic field in (4.4) has a 'cosmological correction term'. [The calculations in King and Pfister (2001) are exact in M and q, the inner shell has no rest mass density, and in particular, rotation is defined by the independent angular momenta J_a and J_R of the two shells, rather than their angular velocities ω_a and ω_R, which are in fact dependent due to the dragging phenomena.] Here, one may interpret the radius R as a measure of 'cosmic' distances (a kind of 'world radius') and the radius a as a measure for terrestrial or laboratory length scales.

In this way, the simplified cosmological model of a collapsed mass shell exhibits an influence of the universe on our local physics (the magnetic field). However, the relative difference between curved and flat space results is, for good or bad, in all conceivable cases beyond measurability (approximately smaller than 10^{-57} if the radius a equals the radius of the Earth). Hence, in the usual Schwarzschild-like coordinate ρ, (4.4) gives the perfectly Machian result that the magnetic field with respect to an inertial observer corotating with the mass shell and well inside (i.e., for all length scales $\rho \ll R$) is exactly the magnetic dipole field of a rotating charged shell as known from classical electrodynamics [together with the corresponding component $B_\vartheta(\rho, \vartheta)$ from flat spacetime electrodynamics]:

$$
B_\rho(\rho, \vartheta) = \begin{cases} \dfrac{2qa^2}{3\rho^3}(\bar{\omega}_a - \bar{\omega}_R)\cos\vartheta & \text{for } a \le \rho \ll R, \\[4mm] \dfrac{2q}{3a}(\bar{\omega}_a - \bar{\omega}_R)\cos\vartheta & \text{for } \rho \le a. \end{cases} \tag{4.5}
$$

To first order in M and q, the magnetic fields already found using different methods by Ehlers and Rindler in Ehlers and Rindler (1970, 1971) on the basis of the conjecture made by Rindler were successfully confirmed in King and Pfister (2001) and shown to be in full agreement with Machian expectations. (In contrast, Ehlers and Rindler have referred to this field as 'Mach-negative or, at best, Mach-neutral'.)

Highly charged two-shell systems display two quite unusual dragging phenomena (King and Pfister 2001). First, as already mentioned, for a Machian interpretation of the dragging effects, the exterior mass shell is often seen as an idealized substitute for part or all of the cosmic masses. For this interpretation to be valid, a minimal condition seems to be that the energy–momentum tensor of this mass shell should satisfy the weak energy condition. If these conditions are violated for some region of the parameter space (a, R, q, M) of the system, this leads to an 'antidragging' phenomenon: the local inertial frames are dragged in the opposite direction to the rotation of the mass shell. In a trivial manner, antidragging already manifests itself for the uncharged, weakly massive Thirring model in the case of a negative shell mass: $d_1 = 4M/3R < 0$ for $M < 0$. The interpretation of this anomalous change of sign of the dragging is as follows: the charged shell in the model has zero mass density and negative 'pressure' (in order to balance the Coulomb repulsion), and therefore nearly violates all energy conditions. It was shown in Pfister and King (2002) that such a shell produces a negative dragging term if it rotates. (The angular momentum J_a of the inner, charged shell may become negative, although ω_a is positive! This again demonstrates that it is the intrinsic angular momentum which causes dragging phenomena, see also Sect. 4.4.)

Second, in the region between the two shells one finds a radially *increasing* dragging function. Such behavior is at first sight barely comprehensible, as typically the gravitomagnetic field outside a rotating body falls off as r^{-3} [compare with (4.2)]. The explanation seems to come from the (positive and nonrotating) electrostatic energy density. Quite generally, the degree of dragging is determined, at least qualitatively, by the ratio between the rotating and the nonrotating mass energy of the whole system (e.g., in the standard Thirring problem, the part of the cosmic masses sitting on a rotating mass shell). If then (for fixed M) a small part of the (rotating) exterior shell is 'replaced' by electrostatic energy, the constant dragging inside the charged shell is reduced. For a fixed value of the radial coordinate $r > a$, the corresponding part of the electrostatic energy density has only a reduced effect because, as stated above, the dragging due to masses quite generally falls off as r^{-3} in their exterior. Therefore, the dragging function increases in the intermediate region $a < r < R$. Furthermore, in the collapse limit of the massive highly charged two-shell system, the important result in Brill and Cohen (1966) that in this limit there results complete dragging of the inertial frames inside the mass shell, extends to the inertial frames inside the mass shell of an electromagnetic Thirring system. One may further conjecture that such perfect dragging is 'universal' in the collapse limit, i.e., irrespective of *all* physical fields inside the (slowly) rotating mass shell. (In the exterior region of the system one finds, as expected, the Kerr–Newman field to first order in ω.)

4.3 Realization of Machian Ideas in Cosmology and in Nature

The dragging phenomena described in Sect. 4.2 confirm that Mach's concept of relativity of inertia, in particular relativity of rotation, is valid in general relativity, at least to some degree, e.g., for small rotation rates. However, since none of the models described hitherto are of a cosmological character, but are based on asymptotically flat spacetimes, they do not, at least not directly, contribute to Mach's hypothesis that inertia is governed by the overall mass distribution in the universe. Before coming to some real cosmological generalizations of the asymptotically flat models, we shall nevertheless argue that these models already display some features which are of interest from a cosmological point of view, and which are at least in qualitative agreement with Mach's hypothesis.

As a first example, we recall the model of a slowly rotating, but heavy mass shell in Brill and Cohen (1966), especially Fig. 2, where it was shown that, in the collapse limit, the interior of the mass shell splits off as a separate 'universe' in which:

> [...] there cannot be a rotation of the local inertial frame in the center relative to the large masses in the universe. In this sense our result explains why the 'fixed stars' are indeed fixed in our inertial frame, and in this sense the result is consistent with Mach's principle.

In the subsequent paper (Cohen and Brill 1968), several concentric mass shells with masses m_i, radii r_i, and angular velocities ω_i are considered, each causing a dragging effect $4m_i\omega_i/3r_i$ at the center. If one now imagines a homogeneous universe consisting of such successive mass shells, all rotating with the same (small) angular velocity ω, then at least in the weak-field approximation, the dragging effects of all shells will superpose linearly, and for a constant mass density of the universe the masses m_i will grow quadratically with r_i. Therefore the contribution of each shell to the overall dragging effect at the center grows linearly with the radius, thereby confirming Mach's conjecture that (Mach 1872) "the share of the nearest masses vanishes in comparison with that of the farthest" (compare Sect. 4.1).

A very interesting and quite important question concerning dragging effects in general relativity and cosmology was first studied in a concrete model in Lindblom and Brill (1974), namely the question of whether inertial effects are instantaneous or retarded. The model system was the free-fall collapse of a slowly rotating dust shell. And the result of a relatively simple calculation was that:

> [the dragging function] $\Omega_{\text{obs}}(t)$ is determined by the 'instantaneous' radius of the shell $R(t)$. That is, as seen from infinity, the inertial frames within the shell rigidly rotate at the angular velocity Ω_{obs}: there are no retardation effects between the shell and the inertia of a gyroscope at its center. This of course does not contradict any physical causality principle, since Ω_- can be considered to be merely the angular velocity of a coordinate system for the interior flat region. However, it is this coordinate system which is most directly related to effects observable from infinity, as explained above. Thus another view, more closely related to Machian ideas, is equally consistent, in which $\Omega_-(t)$ is observable but highly nonlocal, so that a local causality principle does not apply to it. [...] These results fit most simply with the 'spacelike' formulation of Mach's principle. [...] The constraint equations determining N and N^i are purely spacelike equations, and in this sense all Machian effects will be related to the instantaneous values of the dynamical variables.

This view was already anticipated, at least qualitatively, in Wheeler (1964, p. 367):

> The influence of a local increment $\delta\epsilon$, δs^i on the shift function N_i is nonlocal. It shows up in the formalism neither as a retarded effect nor as an advanced effect. Instead, because the analysis deals with an everywhere *spacelike* hypersurface, the influence appears ostensibly as instantaneous.

Below we will document that quite generally, and also in a real cosmological setting, Mach's principle is connected with the time-independent constraint equations of general relativity.

Now we come to attempts to calculate rotational dragging effects in models which replace the asymptotic flatness by more or less realistic cosmological asymptotic or boundary conditions. As far as we are aware, the first such attempt is Lewis (1980). This work starts from the so-called Einstein–Straus vacuole (Einstein and Straus 1945), where a flat interior Minkowski region is continuously connected via a coexpanding spherical mass shell to a spatially closed ($k = 1$) Friedmann dust solution. Axially symmetric perturbations superimposed on this model were calculated in Lewis (1980) to first order in the angular velocity, leading to a dragging factor for the interior inertial frames which depends on the properties of the shell, on the cosmic matter, and on the cosmic time, and "tends to confirm the spirit of Mach's principle, if not the exact definition."

Similarly, in Chamorro (1988), it is pointed out that the slowly rotating mass shells in an asymptotically flat spacetime are 'cosmologizable', and here for all three cases of closed, open, and critically open universes ($k = 1, -1, 0$), but without taking into account the work (Lewis 1980), and without discussing the Machian aspects in any detail. These aspects are then carefully analyzed in Klein (1993), with an invariant definition of the dragging factor by observable quantities (according to Lindblom and Brill 1974), and with a discussion of the relative contributions of the rotating mass shell and the rotating cosmic dust to the overall dragging effect inside the shell. It is also proven there that, in these models, the shell mass exactly equals the mass 'cut out' from the Friedmann universe.

In Lynden-Bell et al. (1995), the authors also begin with Machian effects in slowly rotating mass shells, but then go over to rotational perturbations of closed Friedmann–Robertson–Walker (FRW) cosmologies, and derive an explicit expression for the dragging potential $\omega(r, t)$ as an integral over the angular momentum distribution $J(\leq r, t)$ of the cosmic matter, with an appropriate weight function. And, as announced above, in this general setting, they also come to the conclusion that "the potential $\omega(r, t)$ that governs the rotations of the local inertial frames is instantaneously related to the angular momentum distribution $J(\leq r, t)$", and "that Mach's principle follows from the constraint equations of general relativity, provided that the universe is closed." Besides the detailed mathematical proof of these facts in Lynden-Bell et al. (1995), a simpler and more convincing argument for the instantaneous action of inertia is provided in Katz et al. (1998). A first order rotational perturbation of spherical FRW universes is of purely dipolar character. But general relativity, as a tensorial field theory, does not allow for dipolar causal signals (waves). [This is in contrast to the vectorial theory electrodynamics, as worked out in detail in Katz et al. (1998).] The paper (Lynden-Bell et al. 1995) also contains a simple proof that the angular momentum of any closed universe is necessarily zero (compare also King 1995).

Similar but partly different results have been derived by C. Schmid in a short paper (Schmid 2002), and in two quite extended papers (Schmid 2006) for ($k = 0$) FRW cosmologies, and Schmid (2009) for ($k = \pm 1$) FRW cosmologies. Application of the quite general cosmological perturbation formalism of Bardeen (1980) to rotational perturbations leads to the following essential results (Schmid 2002):

1. The dragging of a gyroscope axis by rotational perturbations beyond the $\dot{H}$ radius (H = Hubble constant) is exponentially suppressed.

2. If the perturbation is a homogeneous rotation inside a radius significantly greater than the $\dot{H}$ radius, then the dragging of the gyroscope axis by the rotational perturbations is exact for any equation of state for cosmological matter.
3. The time evolution of a gyroscope axis exactly follows a specific average of the matter inside the $\dot{H}$ radius for any equation of state.

In this precise sense Mach's principle follows from cosmology with Einstein gravity.

Concerning the first item, it should be noted that this conclusion has been questioned in Bičák et al. (2004): the exponential suppression is only present if the angular velocities of the cosmic matter are prescribed. But it would seem more appropriate to prescribe the angular momenta, for which a conservation law is valid. [Similar conclusions have been drawn for the so-called 'electromagnetic Thirring problem' in King and Pfister (2001).] However, this controversy does not affect the central message of all these papers that rotational perturbations of FRW cosmologies confirm Mach's principle as far as one might wish. In view of the recent observational results (mainly due to the structure of the cosmic background radiation) that our actual universe demands dark matter, or a cosmological constant Λ, it may be instructive to investigate how far the above results concerning Mach's principle generalize to cosmologies with a Λ term.

After this theoretical analysis concerning the realization of Machian ideas in cosmology, we address the question as to whether, and how precisely, the experimentally realized local inertial axes are tied to the overall mass distribution of the universe, i.e., how well Mach's principle works in nature. In a rough and qualitative manner, such a relation was already known to Galileo, Kepler, Newton, and others, and in 1851, L. Foucault demonstrated that a freely suspended pendulum maintains its plane of oscillation with respect to the 'rest system' of the universe during a 24 h revolution of the Earth. The first concrete estimate we have found in the literature for the degree of 'non-rotation' of the local inertial systems against the 'fixed stars' is due to Seeliger (1906), who states "that the practically used empirical astronomical coordinate system does not rotate relative to an inertial system by more than a few arcseconds in a century". From today's perspective one can, however, have doubts as to whether such accuracy was really possible more than 100 years ago.

Nowadays, laser gyros presumably represent the most precise, truly local rotation sensors, and these reach an accuracy of (only) $10^{-8}\omega_E$ of the Earth's angular velocity (Stedman 1997; Schreiber et al. 2008). Obviously, the best rotation sensors today are given by the (less local) Earth-based reference systems of VLBI (Very Long Baseline Interferometers) and the GPS (Global Positioning System), for which the accuracy reaches $10^{-9}\omega_E$ (Kovalevsky et al. 1989). For the dynamical solar reference system, with respect to which the planetary orbits are optimally adjusted to Newton's laws and its well-known relativistic corrections, the precision is $5 \times 10^{-9}\omega_E$ (Kovalevsky et al. 1989). For the galactic reference frame, realized by the HIPPARCOS catalogue (HIgh Precision PARalax COllecting Satellite), the estimate is $7 \times 10^{-8}\omega_E$ (Kovalevsky et al. 1997). This number can presumably be improved by a factor of 100 by the GAIA satellite (Global Astrometric Interferometer for Astrophysics), which was launched on 9 December 2013, and which will map 10^9

stars of the Milky Way over 5 years with a precision of 7×10^{-6} arcsec. None of these experiments have found any deviation from the hypothesis that the (quasi-) local inertial systems are fixed relative to the distant stars and galaxies, and the cosmic background radiation. A number which does not directly measure the (non-) rotation of a local inertial system against the cosmos, but which is nevertheless of interest in this connection, is the vorticity strength of the cosmic background radiation, which is estimated (Kogut et al. 1997) by $\omega/H_0 < 6 \times 10^{-8}$, i.e., less than 10^{-7} 'revolutions' during the whole lifetime of the universe.

We close this section with some wonderful, somewhat provocative quotations from prominent experts, addressing this 'miraculous' connection between local physics and the universe as a whole. (As far as we are aware, the only other such connection results from the 'initial' state—of very low entropy—of the universe, which presumably triggers the time arrow of all our local irreversible processes.) In the textbook Misner et al. (1973, p. 547), we read:

Consider a bit of solid ground near the geographic pole, and a support erected there, and from it hanging a pendulum. Though the sky is cloudy, the observer watches the track of the Foucault pendulum as it slowly turns through 360°. Then the sky clears and, miracle of miracles, the pendulum is found to be swinging all the time on an arc fixed relative to the far-away stars.

S. Weinberg expresses a similar thought experiment in Weinberg (1972, p. 17):

There is a simple experiment that anyone can perform on a starry night, to clarify the issues raised by Mach's principle. First stand still, and let your arms hang loose at your sides. Observe that the stars are more or less unmoving, and that your arms hang more or less straight down. Then pirouette. The stars will seem to rotate around the zenith, and at the same time your arms will be drawn upward by centrifugal force. It would surely be a remarkable coincidence if the inertial frame, in which your arms hung freely, just happened to be the reference frame in which typical stars are at rest, unless there were some interaction between the stars and you that determine your inertial frame.

And S. Hawking writes in Hawking (1969) (reproduced in Misner et al. 1973, p. 938):

The observed isotropy of the microwave background indicates that the universe is rotating very little if at all. [...] This could possibly be regarded as an experimental verification of Mach's principle.

Finally, we hint at a quite alternative quotation from Schücking (1996) which concerns a linear acceleration, and makes it particularly clear that Machian effects are governed by the time-independent constraint equations of general relativity (because the cosmic masses can never react causally to the sudden decision of the car driver to slam on the brakes):

Mach's principles—whatever they may be—will always find their defenders and believers. When one of its promoters, Dennis Sciama, slammed on the brakes of his car, propelling his girlfriend, seated next to him, towards the windshield, she was said to be heard moaning, 'All those distant galaxies'.

4.4 Gravitomagnetism and Its Observational Basis

In the preceding sections, we have given a detailed summary of the general 'dragging of inertial frames' or 'frame-dragging', its history, and its Machian interpretation, and of the origin of inertia: in Einstein's theory of general relativity, not only mass-energy but especially mass-energy currents create curvature and influence spacetime structure. Although the cosmic masses and their motions may not entirely determine the local inertial structure of spacetime, they at least influence it in a way that is explicable in the spirit of Mach. As Ciufolini and Wheeler very vividly put it in a nutshell (Ciufolini and Wheeler 1995, pp. 4 and 399):

> *Inertia here*, in the sense of *local inertial frames*, that is the grip of spacetime here on mass here, is fully defined by the geometry, the curvature, the structure of spacetime here. The geometry here, however, has to fit smoothly to the geometry of the immediate surroundings; those domains, onto their surroundings; and so on, all the way around the great curve of space. Moreover, the geometry in each local region responds in its curvature to the mass in that region. Therefore every bit of momentum–energy, wherever located, makes its influence felt on the geometry of space throughout the whole universe—and felt, thus, on inertia right here.

And put simply:

> In conclusion we may summarize: *mass-energy 'tells' spacetime how to curve and spacetime 'tells' mass-energy how to move.* [...] Therefore, *mass-energy there rules inertia (local inertial frames) here.*

In general, all ten components of the energy–momentum tensor $T_{\mu\nu}$ create curvature and influence the spacetime structure. Gravitation, at least in Einstein's general relativity formulation, is a tensor theory, so the mass density is not the only source of gravitational fields. For all physical systems which have a global time coordinate, as is true, e.g., for standard cosmological models and for stationary and asymptotically flat systems, there exists a well defined separation of the energy–momentum tensor $T^{\mu\nu}$ into the energy density T^{00}, the momentum density T^{0i}, and the stress tensor T^{ik}. In the linear (weak-field) approximation to general relativity, these different sources produce accompanying, well separated gravitational fields.

It has been known for a long time (see below) that the gravitational fields produced by T^{00} and T^{0i} exhibit a close analogy to electromagnetic fields, and therefore the names 'gravitoelectric fields' (produced by T^{00}) and 'gravitomagnetic fields' (produced by T^{0i}) are appropriate (see, e.g., Harris 1991). The latter lead to a phenomenon called gravitomagnetism, a new type of gravitational field or 'force', totally unknown in Newton's theory. (However, in contrast to electromagnetism, pure gravitomagnetic fields do not exist due to the lack of gravitationally neutral matter.) The fields produced by T^{ik}, and which could be called 'gravitotensorial fields', would not appear to have been studied in the same systematic manner (Pfister and Schedel 1987). In the full nonlinear theory of general relativity, the effects of the different sources T^{00}, T^{0i}, T^{ik} are of course mixed, so that a general gravitational field can no longer, even in a fixed coordinate system, be clearly separated into gravitoelectric, gravitomagnetic, and gravitotensorial components.

(In this connection there emerges the interesting question of a possible reduction or even compensation of gravitational attraction through gravitomagnetic effects. Some relativistic models with momentum density T^{0i} and stresses T^{ik} of magnitude comparable to T^{00} are presented below.)

In this section, we first examine gravitomagnetism—or gravitoelectromagnetism (GEM)—on the basis of the linearized Einstein field equations and in close analogy with Maxwell's equations of classical electrodynamics. We then briefly characterize this field intrinsically by its spacetime invariants, and finally comment on the experimental situation.

Historically, the basic equations of (linear) gravitomagnetism were already derived in 1918 by H. Thirring in Thirring (1918). A vague idea of a new gravitomagnetic 'force' already appears in 1896 in a booklet by the Friedlaender brothers in Friedlaender and Friedlaender (1896), and then in the title of an Einstein paper from 1912 (Einstein 1912). More concretely, relativistic field equations for gravitomagnetism appear first in Einstein's talk at the Naturforscher congress 1913 in Vienna (Einstein 1913), based on the preliminary Entwurf theory, where Einstein explicitly says: "The [gravitational] equations correspond largely to those of electrodynamics, [...] up to the sign, and [...] up to a factor 1/2." Thirring was without doubt decisively stimulated by this talk which he quotes in the introduction to his paper. And he explicitly mentions the different sign and a factor 4 in his gravitomagnetic equations in comparison to electromagnetism [as Einstein did in his talk (Einstein 1913)]. Although he presents no physical interpretation of these differences, it was no doubt clear to him that the different sign comes from the fact that all (positive!) masses attract each other, whereas charges of equal sign repel each other. Not so evident is whether Thirring was aware of the fact that the factor 4 results from the tensorial character (spin 2) of general relativity, in contrast to the vectorial theory electrodynamics (spin 1 of the photons).

Now, the field equations, derived from a linear perturbation of Minkowski space-time given by $g_{\mu\nu} = \eta_{\mu\nu} + h_{\mu\nu}$ with $|h_{\mu\nu}| \ll 1$, where $\eta_{\mu\nu}$ is the flat Minkowski (background) metric, and subject to certain 'simplifying' gauge conditions, are the following (groups of homogeneous and inhomogenous) Maxwell-type equations [in a notation adapted from Harris (1991) and Ohanian and Ruffini (2013)]:

$$\nabla \cdot \boldsymbol{H} = 0 , \tag{4.6}$$

$$\nabla \times \boldsymbol{g} = 0 , \tag{4.7}$$

$$\nabla \cdot \boldsymbol{g} = -4\pi(2T^{00} - T) , \tag{4.8}$$

$$(\nabla \times \boldsymbol{H})^i = -16\pi T^{0i} , \tag{4.9}$$

where **H** is the 'gravitomagnetic field' and **g** is the standard Newtonian gravitational or 'gravitoelectric' field. Likewise, the source term $2T^{00} - T$ in (4.8) may analogously be called the 'gravitational charge density', and the source term T^{0i} in (4.9) the 'gravitational current density'. (Hence, in the case of an energy–momentum tensor of, e.g., a perfect fluid, there are also contributions to **g** due to the pressure.)

For stressless matter (so-called dust), these source terms reduce to the mass energy density ϱ_m and the mass current density $\mathbf{j}_m = \varrho_m \mathbf{v}$.

The gravitational fields are related to the respective potentials by $\mathbf{H} = \nabla \times \boldsymbol{h}$, where the so-called gravitomagnetic potential $\mathbf{h} = (h_{01}, h_{02}, h_{03})$ is the off-diagonal $(0i)$-component of the metric, and to the Newtonian potential $h_{00} = h_{11} = h_{22} = h_{33} = 2\Phi$ by $\mathbf{g} = -\nabla\Phi$. (In the literature, these two gravitational fields are occasionally denoted by $\mathbf{B_g}$ and $\mathbf{E_g}$, with the potentials $\mathbf{A_g}$ and Φ_g.) The (gauge-dependent) equations of gravitomagnetism (4.6)–(4.9) are restricted to the weak-field approximation of general relativity, to *stationary* (i.e., effectively time-independent) gravity fields, and to slowly moving source masses and test masses (in the measuring device), something that would seem to be realistic for all foreseeable experiments measuring gravitomagnetism (see below).

It should, however, be mentioned that, as in electromagnetism, the strength of the gravitomagnetic field in relation to the gravitoelectric (quasi-Newtonian) field depends to a large extent on the velocity of the observer and that, in the literature, there are some misunderstandings about which effects are real gravitomagnetic effects and which are not. There are even misunderstandings about the validity of this linear approach to GEM. Within a Maxwell-type, covariant formulation of GEM based on 'tidal tensors' and for strong gravity fields, Costa and Herdeiro (2008) showed that the linear GEM formulation, with its noncovariant and gauge-dependent fields, are valid only for weak, stationary, i.e., time-independent fields, and slowly moving sources. In particular, there is no gravitational analogue to Faraday's law of induction. [A careful mathematical analysis of the role of gauge transformation in this context is given by Clark and Tucker (2000) with the same result concerning the time dependence of solutions to different GEM equations.]

Before providing solutions to (4.6)–(4.9), we come back to the above-mentioned counteraction between gravitational attraction and gravitomagnetic repulsion for two relativistically rotating bodies. In Pfister and Schedel (1987), it has been shown in the weak-field approximation of general relativity that, for two infinitely thin spherical shells of matter, corotating about the same axis (angular velocity ω), gravitomagnetic repulsion can partly compensate for the attraction. In the extreme limit of infinitely high 'multipolarity' of the mass density, the shells degenerate to mass rings. If the distance between the two bodies vanishes, and if we take the limit $v/c = \omega R/c \to 1$ (but only in this unphysical limit!), gravitational attraction exactly balances gravitomagnetic repulsion. In the case of two aligned rotating black holes with parallel spins, where the 'spin–spin interaction' generating repulsive effects may compensate gravitational attraction, Neugebauer and Hennig (2009) and Hennig and Neugebauer (2011) showed non-existence for these two-black-hole configurations.

For a rotationally accelerated central mass (linearly accelerated heavy masses and their corresponding gravitational fields play practically no role in any relativistic situation), the corresponding mass-energy currents create a gravitomagnetic dipole field, just as in electrodynamics, e.g., a rotating charged sphere (or the current due to the rotation of the shell) is the source of a magnetic (dipolar) field (see, e.g., Ciufolini 1994; Ciufolini and Wheeler 1995). In this case, the field equations of general relativity in the weak-field and slow-motion approximation take a form very close to Maxwell's equation for a stationary electric current distribution $\mathbf{j}_q = \varrho_q \mathbf{v}$ in the Coulomb gauge $\Delta \mathbf{A} = -4\pi \mathbf{j}_q$, where $\mathbf{A}$ is the (electromagnetic) vector potential and $\mathbf{B} = \nabla \times A$ is the corresponding magnetic field. Far from the source, the leading order of the potential in a formal power series in $1/r$ is a magnetic dipole vector potential $\mathbf{A}(\mathbf{r}) = (\mathbf{m} \times \mathbf{r})/r^3$, which has a radial fall-off behavior $\sim 1/r^2$, and where $\mathbf{m}$ is the magnetic moment of the stationary current distribution. Hence, the magnetic dipole field reads

$$\mathbf{B}(\mathbf{r}) = \frac{1}{r^3}\left[-\mathbf{m} + \frac{3\mathbf{r}}{r^2}(\mathbf{m} \cdot \mathbf{r})\right] \sim 1/r^3 .$$

Similarly, and by *formal* analogy with Ampere's law, in the weak-field and slow-motion limit, subject to the Lorenz gauge, and for a stationary mass-energy current distribution $\mathbf{j}_m = \rho_m \mathbf{v}$, Einstein's field equation $\Delta \mathbf{h} = 16\pi \mathbf{j}_m$ may be solved for the gravitomagnetic potential $\mathbf{h}$ and field $\mathbf{H}$. Again, far from the gravitating system in the asymptotic regime, the leading term $\sim 1/r^2$ is $\mathbf{h}(\mathbf{r}) = -2(\mathbf{J} \times \mathbf{r})/r^3$, where $\mathbf{J}$ is the intrinsic angular momentum of the central mass. The vector potential $\mathbf{h}$ now defines the gravitomagnetic field:

$$\mathbf{H}(\mathbf{r}) = \nabla \times \mathbf{h} = \frac{2}{r^3}\left[\mathbf{J} - \frac{3\mathbf{r}}{r^2}(\mathbf{J} \cdot \mathbf{r})\right] \sim 1/r^3 . \tag{4.10}$$

Noting the formal equivalence of the magnetic moment $\mathbf{m}$ and the angular momentum $\mathbf{J}$, the gravitomagnetic field $\mathbf{H}$ looks quite similar to the magnetic dipole field $\mathbf{B}$, except, as already mentioned, for the minus sign (as gravity is always attractive) and a factor 2 (reflecting the different spin of the gravitational and electromagnetic interactions). Despite the fact that this new gravitational field has been derived in the approximations made above, it has to be stressed that the phenomenon of gravitomagnetism is *not* restricted to weak gravitational fields and slowly moving objects (see the invariant characterization of gravitomagnetism below)!

The geodesic equation on stationary spacetimes, exhibiting weak gravitational fields, due to slowly moving and stationary mass currents reads

$$m\ddot{\mathbf{r}} = m(\mathbf{g} + \mathbf{v} \times \mathbf{H}) , \qquad \mathbf{g}(\mathbf{r}) = -\frac{M}{r^2}\frac{\mathbf{r}}{r} ,$$

where $\mathbf{g}$ is the standard Newtonian acceleration. This equation of motion for a test particle with mass m is once again strikingly similar to the Lorentz force field of a moving charge q in classical electrodynamics, viz., $m\ddot{\mathbf{r}} = q[\mathbf{E} + (\mathbf{v}/c) \times \mathbf{B}]$. The Lorentz force exerts a torque $\tau = \mathbf{m} \times \mathbf{B}$ on a magnetic dipole moment, and this, by analogy, transfers to a torque on a test gyroscope with spin $\mathbf{S}$ in general relativity. Then from the equation for parallel transport of a spin vector S^μ, one gets

$$\tau = \frac{1}{2}\mathbf{S} \times \mathbf{H} = \frac{d\mathbf{S}}{dt} \equiv \mathbf{\Omega} \times \mathbf{S} ,$$

with an angular velocity

$$\boldsymbol{\Omega} = -\frac{1}{2}\mathbf{H} = -\frac{1}{r^3}\left[\mathbf{J} - \frac{3\mathbf{r}}{r^2}(\mathbf{J}\cdot\mathbf{r})\right] ,\qquad(4.11)$$

with respect to an observer in an asymptotic inertial system. This is the standard formula quoted as 'dragging of inertial frames', where the axes of the local inertial systems are defined operationally by the direction of the spin of test gyroscopes. For a gyro located at the north pole of the (hypothetical exactly spherically symmetric) Earth, one can estimate the precession rate to be of the order of

$$\Omega_{\mathrm{gyro}}(\text{north pole}) = \frac{4}{5}\frac{GM}{c^2 R}\omega_{\mathrm{E}} \approx 5.5 \times 10^{-10}\omega_{\mathrm{E}} .$$

Now, the gravitomagnetic field $\mathbf{H}$ is linked to a 'force'

$$\mathbf{F} = \left(\frac{1}{2}\mathbf{S}\cdot\nabla\right)\mathbf{H}$$

on the gyro's spin. This force field is sometimes called the 'gravitomagnetic force'. Although in Newtonian theory the concept of a force is of central importance and crucial for the dynamics of bodies, this concept, so extremely successful in our everyday life, is finally fading out and even breaks down in general relativity. In particular, inertial and gravitational forces are unified, and gravity is no longer a force in the Newtonian sense. The fundamental field is now the spacetime metric $g_{\mu\nu}$, from which we deduce the affine connection $\Gamma^{\lambda}_{\mu\nu}$ characterizing the inertial structure of spacetime. For these reasons, we avoid the term 'gravitomagnetic force' in the general relativistic spacetime picture. [However, for some illustrative thought experiments on gravitational 'forces' in general relativity, well definable for static and stationary spacetimes, see Lynden-Bell and Katz (2014).]

From (4.11), another well-known formula may be derived, namely the Lense–Thirring precession (Lense and Thirring 1918) of a test particle orbiting a central mass with angular momentum $\mathbf{J}$. The Lense–Thirring nodal precession describes the secular rate of change of the orientation of the orbital plane and the longitude of the nodes:

$$\boldsymbol{\Omega}^{\text{Lense–Thirring}} = \frac{2\mathbf{J}}{a^3(1-e^2)^{3/2}} .\qquad(4.12)$$

Here, the orbital parameters of the 'test' object (e.g., an artificial satellite, the Moon orbiting the Earth, etc.) are the semimajor axis a and the eccentricity e. For a circular ($e = 0$) orbit (radius r) of the test particle, (4.12) simplifies to $\boldsymbol{\Omega}^{\text{Lense–Thirring}} = 2\mathbf{J}/r^3$.

To characterize gravitomagnetism without recourse to any frame attached to, or any coordinate system labelling spacetime, we refer this new gravitational field intrinsically to spacetime invariants. The presentation given here follows closely the method proposed in Ciufolini (1994), Ciufolini (1995, pp. 386–402), and Ciufolini and Wheeler (1995, Sect. 6.11). As for the field equations for the gravitomagnetic potential **h** and field **H**, we motivate this characterization by its formal analogy with classical and special relativistic electrodynamics. In Maxwell's theory on flat Minkowski spacetime, there exist two different scalar Lorentz invariants, i.e., algebraic combinations of the electric and the magnetic field, **E** and **B**, which are the coefficients of the characteristic polynomial of the skewsymmetric electromagnetic field tensor $F^{\mu\nu}$. First, the invariant

$$I_1 = -\frac{1}{2} F_{\alpha\beta} F^{\alpha\beta} = |\mathbf{E}^2| - |\mathbf{B}^2| \,,$$

and second, the pseudoinvariant

$$I_2 = \frac{1}{4} F_{\alpha\beta} {}^\star F^{\alpha\beta} = \mathbf{E} \cdot \mathbf{B} \,.$$

Here ${}^\star F^{\alpha\beta} = \varepsilon^{\alpha\beta\mu\nu} F_{\mu\nu}/2$, where $\varepsilon^{\alpha\beta\mu\nu}$ is the Levi-Civita pseudotensor, and $\star$ the duality operation. Now, in the rest frame of a charge q (without a magnetic dipole moment **m**), we have only a nonvanishing electric field (the electrostatic Coulomb field), and therefore $I_1 \neq 0$ and $I_2 = 0$. And this even in an inertial system moving with velocity **v** with respect to the rest frame, and where both $\mathbf{E} \neq \mathbf{0}$ and $\mathbf{B} \neq \mathbf{0}$ (the magnetic field according to the transformation property of the field tensor $\hat{F}^{\mu}_{\nu} = \Lambda^{\mu}_{\alpha} \Lambda^{\beta}_{\nu} F^{\alpha}_{\beta}$). It is only in a rest frame with a charge *and* a magnetic dipole moment that $I_2 \neq 0$ in any case, i.e., independently either of the chosen Lorentz frame or due to a mere coordinate transformation.

In Einstein's theory of general relativity on curved spacetime, there are invariants of the Riemann tensor $R_{\mu\nu\lambda\kappa}$ as well, e.g., the Ricci scalar R vanishes in flat spacetime, again irrespective of the frame attached to, or the coordinates labelling, spacetime. [For a complete classification of the 14 independent curvature scalar invariants, of which only four are independent in empty space, see Petrov (1969, Sect. 21).] A further invariant is the Kretschmann invariant $K_1 = R_{\mu\nu\lambda\kappa} R^{\mu\nu\lambda\kappa}$, which for spacetimes characterized by mass M and angular momentum J (like the Kerr black hole family) is a function $f(M/r^3, J/r^4)$ with the leading term proportional to the mass. Therefore, the Kretschmann invariant is nonvanishing even if the angular momentum is zero. Now, mass-energy currents influence spacetime and create curvature, so gravitomagnetism may not be intrinsically determined by the Kretschmann scalar. As in the electromagnetic case described above, there is an invariant analogous to I_2, the pseudotensor

$$K_2 = [{}^\star R]_{\mu\nu\lambda\kappa} R^{\mu\nu\lambda\kappa} = \frac{1}{2} \varepsilon^{\alpha\beta\mu\nu} R^{\rho\sigma}_{\mu\nu} R_{\alpha\beta\rho\sigma}$$

or Chern–Pontryagin invariant built from the Riemann tensor and the Levi-Civita pseudotensor, which characterizes the gravitomagnetic field *and* the spacetime structure (in contrast to I_2, which characterizes only the field $F^{\mu\nu}$ and not the Minkowski spacetime $\eta_{\mu\nu}$). This invariant (and Kretschmann's invariant) have already been given by Matte (1953) and Petrov (1969), but without any physical discussion of the quantity or any reference to gravitomagnetism. [For an overview of second order scalar invariants of the Riemann tensor, i.e., the Kretschmann invariant K_1, Chern–Pontryagin invariant K_2, and Euler invariant $K_3 = [^\star R\,^\star]_{\mu\nu\lambda\kappa} R^{\mu\nu\lambda\kappa}$, in the framework of the GEM formalism, see Cherubini et al. (2002) and also Costa and Herdeiro (2008).]

Now, the angular momentum J changes sign by a time reflection, as can be inferred from the asymptotic behavior of the $g_{\phi t}$ component of the metric of a stationary solution to Einstein's field equations (see, e.g., Misner et al. 1973, Sect. 19.3). Since the pseudotensor K_2 changes sign under the coordinate transformation $t \to -t$ as well, the pseudotensor is proportional to the intrinsic angular momentum and an odd function of the intrinsic mass-energy currents. Asymptotically and in the weak-field limit, one gets for any stationary solution $K_2 \sim (JM/r^7)\cos\theta$, indicating a gravitomagnetic contribution to spacetime geometry and curvature. Put another way, the Chern–Pontryagin invariant K_2 indicates the existence and presence of a gravitomagnetic field **H**, just as the electromagnetic invariant I_2 determines the existence and presence of a magnetic field **B**, a satisfactory concept which rests entirely on spacetime invariants and not on coordinate dependent 'magnetic' components $g_{0\mu}$ of the metric, or $R_{\mu 0 \nu \sigma}$ of the Riemann tensor (which can always be made nonzero by local Lorentz transformations according to $\hat{g}_{0i} = \Lambda_0^\alpha \Lambda_i^\beta g_{\alpha\beta}$ and $\hat{R}_{i0jk} = \Lambda_i^\alpha \Lambda_0^\beta \Lambda_j^\mu \Lambda_k^\nu R_{\alpha\beta\mu\nu}$, respectively).

In the last part of this section, we give an overview of the experimental status of the measurement of gravitomagnetism. Since most of the tests are Earth-based experiments or satellite missions in Earth orbit, we first estimate the accuracy which has to be achieved here (Pfister 2012, 2014). For laboratories on Earth and for satellites there is a factor $M_E/R_E \approx 10^{-9}$ for any deviations from Newtonian gravity. For rotational accelerations another factor $\omega_E R_E/c \approx 10^{-6}$ accumulates with this to yield a factor 10^{-15} for any gravitomagnetic field in comparison to Newtonian gravity. Since, in contrast to electromagnets, there exist no gravitomagnetic materials in nature, there is typically another factor $v/c \le 10^{-5}$ from the velocity v of the rotating parts of the measuring device (except where these are photons or neutrinos). Presumably, the resulting demand for a total precision of 10^{-20} will not be fulfilled by any laboratory experiment in the foreseeable future, although some groups are making attempts in this direction, e.g., with an underground multi-ring laser gyroscope (Bosi et al. 2011), or with Bose–Einstein condensates in a drop tower (van Zoest et al. 2010). For neutron stars, pulsars, and black holes, the figures for M/R and v/c are of course much more favorable, but there are usually competing, poorly understood processes near these systems. Furthermore, near neutron stars, there are of course no laboratories or other precisely measurable test systems, and for greater distances from the star, the gravitomagnetic field falls off as r^{-3}, as any dipole field does. Hence, even for these astrophysical systems, there is not much hope for an unambiguous measurement of gravitomagnetism in the near future (Stella and Possenti 2009).

Now, such experiments are generally of the following kind (Pfister 2004). Imagine a physicist performing experiments in a closed local laboratory, and in particular determining the local inertial systems relative to the laboratory walls, using various kinds of physical phenomena, e.g., Foucault pendulums, gyros, Sagnactype experiments, and other experiments based on electromagnetic interactions, etc., but of the highest possible precision. In all cases the result (which is in part a consequence of the equivalence principle) is the same (modulo Galilean and Lorentz transformations)! Having done this, he opens the 'windows' of his laboratory and looks up at the sky and the universe. He will (hopefully) be entirely puzzled (MTWs 'miracle of miracles', mentioned in the previous section) to find that the local inertial systems are in fact non-accelerated, in particular non-rotating, with respect to the 'cosmic rest system', i.e., the faraway stars, galaxies, quasars, and the cosmic background radiation. Today, this observational fact seems to be the most significant hint of the influence of the cosmos as a whole on our local physics. As stated in detail in Sect. 4.3, this 'cosmic coincidence', i.e., the non-rotation of the local inertial systems relative to the 'fixed' stars (which has no explanation at all in Newtonian gravity and Newtonian cosmology), has been tested experimentally to an accuracy of 10^{-9} of the Earth's rotation.

All other rotationally accelerated 'Machian experiments' try, either directly or indirectly, to measure the tiny Lense–Thirring effect. These experiments which aim to measure the gravitomagnetic field **H** take place in gravity's weak-field regime (mainly in the Solar System, and in particular in the weak gravitational field of the rotating Earth) and rest on the slow-motion approximation of the test particle (e.g., an orbiting artificial satellite). Historically, the first proposal was made by L. Schiff in Schiff (1960), who derived the equations of motion for the spin **S** of a test particle [see (4.10)]:

$$
\frac{d\mathbf{S}}{dt} = \boldsymbol{\Omega} \times \mathbf{S}, \quad \text{with } \boldsymbol{\Omega} = \frac{1}{2}\mathbf{F} \times \mathbf{v} + \frac{3M}{2r^3}(\mathbf{r} \times \mathbf{v}) + \frac{I}{r^3}\left[\frac{3\mathbf{r}}{r^2}(\boldsymbol{\omega} \cdot \mathbf{r}) - \boldsymbol{\omega}\right],
$$
(4.13)

where **F** is any external, non-gravitational force (vanishing for geodetic motion), and M, I, and ω are the Earth's mass, moment of inertia, and angular velocity, respectively, with **r** and **v** the distance from the Earth's center and the velocity of the gyroscope, respectively. The second term of $\boldsymbol{\Omega}$ in (4.13) is called the geodetic or deSitter precession $\boldsymbol{\Omega}^{\text{deSitter}}$, and was first measured with an accuracy of 2 % by lunar laser ranging of the Earth–Moon system in the gravitational field of the Sun (Shapiro et al. 1988). [At the present time, lunar laser ranging can achieve an accuracy of about 6×10^{-3} for the deSitter effect, see references in Ciufolini (2007).] The last term characterizes the Lense–Thirring precession $\boldsymbol{\Omega}^{\text{Lense–Thirring}}$.

A rather high-technology satellite project to measure both precession effects, pushing technologies to an extreme in many places (and costing US\$700 million), has been developed in Stanford at the Hansen Laboratory since the 1960s by

William Fairbank, Francis Everitt, and coworkers. This is Gravity Probe B. (In principle, satellite missions, apart from the fact that they are automatically quite clean systems in high vacuum and at low temperature, have the advantage that they can run for years and over very many revolutions of the satellite, in this way partly compensating for the challenging precision demand of 10^{-20} estimated above.) For a detailed overview of this space mission see, e.g., Everitt et al. (2001) and Overduin (2010). The NASA satellite, launched on 20 April 2004, orbits Earth on a polar, 'low' altitude orbit of about 650 km (because of the $1/r^3$ radial fall-off behavior of $\Omega^{\text{Lense–Thirring}}$), in order to eliminate higher multipole moments of the Earth gravitational field, and to separate in direction both precession effects in an optimal way.

The core of the experiment consisted of four superconducting gyros, 'free' of drag to the level of 10^{-10} g in 2,440l of superfluid helium at 1.8 K. In detail, each gyro is a quartz sphere of approximately 4 cm in diameter, coated with a superconducting niobium layer, and homogeneous and spherical to within 10^{-6} (probably the most spherical objects ever manufactured!), and rotating in a quartz housing at a frequency of about 100 Hz. The satellite is carried along with the freely flying quartz spheres in order to get $\mathbf{F} \equiv \mathbf{0}$ in (4.13). The spins of the gyros and their rates of change of direction with respect to the fixed star Rigel, and the respective London magnetic moments were read out by a cryogenic SQUID magnetometer. With it, the Earth's and other magnetic fields had to be shielded to better than 10^{-6} G.

The data-collecting phase of the mission lasted nearly a year and data analysis of Gravity Probe B showed unforeseen effects and unexpected complications due to problems with electric charges on the gyroscopes. These electrostatic contact potentials originated because, while both the gyro rotors and quartz housings achieved almost perfect mechanical sphericity, they were not quite spherical electrically, with the consequence of additional, anomalous Newtonian torques on the spins. In summary, after more than 35 years of preparation and after 5 years of sophisticated data analysis and error corrections, Gravity Probe B has succeeded in directly measuring both precession terms. The theoretically predicted geodetic precession $\Omega^{\text{deSitter}} \approx 6.602$ arcsec/year has been observed at 0.28 %, and the Lense–Thirring precession $\Omega^{\text{Lense–Thirring}} \approx 42$ marcsec/year with an accuracy of approximately 19 % (Everitt et al. 2011). Initially, the space mission was planned with intended accuracies of about 2×10^{-5} % and 0.3 %, respectively.

An alternative, rather sophisticated, and quite 'inexpensive' experiment compared to GP-B, based on existing technologies was proposed in the period 1984–1986 by I. Ciufolini to measure the Lense–Thirring effect (Ciufolini 1986) [for a detailed overview see, e.g., Ciufolini and Wheeler (1995, Sects. 6.7 and 6.8)]. Unlike GP-B, LAGEOS (LAser-ranged GEOdynamic Satellite) is a 'high'-altitude satellite mission primarily designed to detect plate tectonic motions, natural resources, the rotation and higher multipoles of the Earth, etc. Since 1976, a small satellite (only 30 cm in diameter, 400 kg), which is laser-tracked to an accuracy of the order of 10^{-8} and better, has been orbiting at about 5,900 km. The Lense–Thirring effect results in a precession of the nodal lines (the intersection of the satellite orbital plane with the equatorial plane of the Earth) of approximately $\Omega_{\text{LAGEOS}}^{\text{Lense–Thirring}} = 2J/a^3(1-e^2)^{3/2} = 0.031$ marcsec/year, where J, a, and e are the angular momentum, semimajor axis, and orbital eccentricity of the satellite, respectively [see (4.12)]. However, this precession is completely masked by the 'classical' precession of about $\Omega_{\text{LAGEOS}}^{\text{classical}} = 126°$/year, due to the even zonal harmonic coefficients J_{2n} of the Earth.

Ciufolini's key idea was to launch a second LAGEOS satellite with an inclination $I_2 = 180° - I_1$ and otherwise equal orbital parameters, which eliminates the classical precession of the nodal lines ($\Omega_{I_1}^{\text{classical}} = -\Omega_{I_2}^{\text{classical}}$) but doubles the Lense–Thirring effect. LAGEOS II was launched in 1992 at an altitude of approximately 5,800 km with a 'suboptimal' inclination of $I_2 = 162.5° - I_1$. Data analysis over 11 years of laser ranging of the nodal precession of the two satellites finally culminated in a quite conservative estimate of the total uncertainty of about 10 % of the Lense–Thirring effect (Ciufolini and Pavlis 2004; Ciufolini 2007). Most of the error sources, namely the periodic perturbations, are averaged out, while the Lense–Thirring drag is cumulative. The main progress was accomplished by a careful error analysis and in particular by the use of an improved model of the terrestrial gravitational field. The space mission GRACE (Gravity Recovery And Climate Experiment), launched in 2002, provided an enhanced Earth gravity model which included the even zonal coefficients J_{2n} (the main contribution to the uncertainty comes from the one due to the Earth's axially symmetric departure from sphericity δJ_4, and from the uncertainty δJ_2 in the Earth's quadrupole moment, describing the Earth's oblateness) (see, e.g., Ciufolini et al. 2010).

An improvement of about one order of magnitude (i.e., an achieved accuracy of approximately 1 %) of the measurement of the Lense–Thirring effect due to the Earth's gravitomagnetic field is expected from the recently (2012) launched, laser-tracked satellite LARES (LAser RElativity Satellite). Currently, this small spacecraft approximates the behaviour of a 'test-particle' at best, i.e., it shows the smallest deviations from geodesic motion of any artificial satellite in free fall orbit [its residual mean acceleration away from geodesic motion is less than 0.5×10^{-12} m/s^2 (Ciufolini et al. 2012)]. Due to the Ehlers–Geroch theorem (Ehlers and Geroch 2004), which asserts that small (but extended) massive bodies move on near-geodesics (if the Einstein tensor satisfies the dominant energy condition), this fact is in a way crucial for tests of general relativity based on test-particle assumptions.

Finally, the role of gravitomagnetism may also be inferred from another, not yet directly measured phenomenon of Einstein's gravitational theory of general relativity, totally unknown to its Newtonian counterpart, the existence of gravitational waves (as detected indirectly by the pulsar PSR 1913+16). If gravitational waves can be analyzed in detail in the future, this will also be an indirect test for gravitomagnetism, because gravitational waves have equal gravitoelectric and gravitomagnetic contributions (just as in the case of electromagnetism, where electromagnetic waves would be barely conceivable without the existence of magnetism).

References

Barbour, J., Pfister, H. (eds.): Mach's Principle—From Newton's Bucket to Quantum Gravity. Birkhäuser, Boston (1995)

Bardeen, J.M.: Gauge-invariant cosmological perturbations. Phys. Rev. D **22**, 1882–1905 (1980)

Bass, L., Pirani, F.: On the gravitational effects of distant rotating masses. Philos. Mag. **46**, 850–856 (1955)

Bičák, J., Lynden-Bell, D., Katz, J.: Do rotations beyond the cosmological horizon affect the local inertial frame? Phys. Rev. D **69**, 064011 (2004)

Bosi, F., et al.: Measuring gravitomagnetic effects by a multi-ring-laser gyroscope. Phys. Rev. D **84**, 122002 (2011)

Brans, C.H.: Mach's principle and the locally measured gravitational constant in general relativity. Phys. Rev. **125**, 388–396 (1962)

Brill, D.R., Cohen, J.M.: Rotating masses and their effect on inertial frames. Phys. Rev. **143**, 1011–1015 (1966)

Chamorro, A.: A Kerr cavity with a small rotation parameter embedded in Friedmann universes. Gen. Relativ. Gravit. **20**, 1309–1323 (1988)

Cherubini, C., et al.: Second order scalar invariants of the Riemann tensor: applications to black hole spacetimes. Int. J. Mod. Phys. D **11**, 827–841 (2002)

Ciufolini, I.: Measurement of the Lense–Thirring drag on high-altitude, laser-ranged artificial satellites. Phys. Rev. Lett. **56**, 278–281 (1986)

Ciufolini, I.: Gravitomagnetism and status of the LAGEOS III experiment. Class. Quantum Gravit. **11**, A73–A81 (1994)

Ciufolini, I.: Dragging of inertial frames, gravitomagnetism, and Mach's principle. In: Barbour, J.B., Pfister, H. (eds.) Mach's Principle—From Newton's Bucket to Quantum Gravity. Birkhäuser, Boston, 386–402 (1995)

Ciufolini, I.: Dragging of inertial frames. Nature **449**, 41–47 (2007)

Ciufolini, I., Pavlis, E.C.: A confirmation of the general relativistic prediction of the Lense–Thirring effect. Nat. Lett. **431**, 958–960 (2004)

Ciufolini, I., Wheeler, J.A.: Gravitation and Inertia. Princeton University Press, New Jersey (1995)

Ciufolini, I., Pavlis, E.C., Ries, J.C., Koenig, R., Sindoni, G., Paolozzi, A., Neumayer, H.: Test of gravitomagnetism with the LAGEOS and GRACE satellites. In: Ciufolini, I., Matzner, R. (eds.) General Relativity and John Archibald Wheeler, pp. 371–434. Springer, Berlin (2010)

Ciufolini, I., Paolozzi, A., Pavlis, E., Ries, J., Gurzadyan, V.G., Koenig, R., Matzner, R., Penrose, R. Sindoni, G.: Testing general relativity and gravitational physics using the LARES satellite. Eur. Phys. J. Plus **127**, 1–7 (2012)

Clark, S.J., Tucker, R.W.: Gauge symmetry and gravito-electromagnetism. Class. Quantum Gravit. **17**, 4125–4157 (2000)

Cohen, J.M.: Electromagnetic fields and rotating masses. Phys. Rev. **148**, 1264–1268 (1966)

Cohen, J.M., Brill, D.R.: Further examples of 'Machian' effects of rotating bodies in general relativity. Nuovo Cimento B **56**, 209–219 (1968)

Costa, L., Herdeiro, C.: Gravitoelectromagnetic analogy based on tidal tensors. Phys. Rev. D **78**, 024021 (2008)

De La Cruz, V., Israel, W.: Spinning shell as a source of the Kerr metric. Phys. Rev. **170**, 1187–1192 (1968)

de Sitter, W.: On the relativity of inertia. Remarks concerning Einstein's latest hypothesis. Konink. Acad. Wetensch. Amsterdam Proc. Sec. Sci. **19**, 1217–1225 (1917)

Ehlers, J., Geroch, R.: Equation of motion of small bodies in relativity. Ann. Phys. **309**, 232–236 (2004)

Ehlers, J., Rindler, W.: A gravitationally induced (Machian) magnetic field. Phys. Lett. **32A**, 257–258 (1970)

Ehlers, J., Rindler, W.: An electromagnetic Thirring problem. Phys. Rev. D **4**, 3543–3552 (1971)

Einstein, A.: Gibt es eine Gravitationswirkung, die der elektrodynamischen Induktionswirkung analog ist?, Vierteljahrschrift für gerichtliche Medizin und öffentliches Sanitätswesen **44** 37–40 (1912). See also Einstein, A.: In: Klein, M.J., et al. (eds.) The Collected Papers of Albert Einstein, vol. 4, pp. 174–179. Princeton University Press, Princeton (1995)

Einstein, A.: Zum gegenwärtigen Stande des Gravitationsproblems. Phys. Zs. **14**, 1249–1266 (1913)

Einstein, A.: Näherungsweise Integration der Feldgleichungen der Gravitation, pp. 688–696. Sitzb. Preuss. Akad. Wiss., Berlin (1916)

Einstein, A.: Kosmologische Betrachtungen zur allgemeinen Relativitätstheorie, pp. 142–152. Sitzb. Preuss. Akad. Wiss., Berlin (1917)

Einstein, A.: Prinzipielles zur allgemeinen Relativitätstheorie. Ann. Phys. Lpz. **55**, 241–244 (1918)

Einstein, A.: The Meaning of Relativity. Methuen, London (1922)

Einstein, A., Straus, E.G.: The influence of the expansion of space on the gravitational fields surrounding the individual stars. Rev. Mod. Phys. **17**, 120–124 (1945)

Everitt, C.W.F., et al.: Gravity probe B: countdown to launch. In: Lämmerzahl, C., Everitt, C.W.F., Hehl, F.W. (eds.) Gyros, Clocks, Interferometers: Testing Relativistic Gravity in Space. Lecture Notes in Physics, vol. 562, pp. 52–82. Springer, Berlin (2001)

Everitt, C.W.F., et al.: Gravity probe B: final results of a space experiment to test general relativity. Phys. Rev. Lett. **106**, 221101 (2011)

Farhoosh, H., Zimmermann, R.L.:Killing horizons and dragging of the inertial frame about a uniformly accelerating particle. Phys. Rev. D **21**, 317–327 (1980)

Föppl, A.: Über einen Kreiselversuch zur Messung der Umdrehungsgeschwindigkeit der Erde. Sitzb. Bayer. Akad. Wiss. math.-nat. Klasse **34**, 5–28 (1904a)

Föppl, A.: Über absolute und relative Bewegung. Sitzb. Bayer. Akad. Wiss. math.-nat. Klasse **34**, 383–395 (1904b). English translation In: Renn, J., et al. (eds.) The Genesis of General Relativity, vol. 3, pp. 145–152. Springer, Dordrecht (2007)

Frauendiener, J., Klein, C.: Exact relativistic treatment of stationary counterrotating dust disks: physical properties. Phys. Rev. D **63**, 084025 (2001)

Friedlaender, B., Friedlaender, I.: Absolute Oder Relative Bewegung? Simion, Berlin (1896). English translation In: Renn, J., et al. (eds.) The Genesis of General Relativity, vol. 3, pp. 127–144. Springer, Dordrecht (2007)

Gödel, K.: An example of a new type of cosmological solution of Einstein's field equations of gravitation. Rev. Mod. Phys. **21**, 447–450 (1949)

Grøn, Ø., Eriksen, E.: Translational inertial dragging. Gen. Relativ. Gravit. **21**, 105–124 (1980)

Hansen, R.O., Winicour, J.: Killing inequalities for relativistically rotating fluids. J. Math. Phys. **16**, 804–808 (1975)

Hansen, R.O., Winicour, J.: Killing inequalities for relativistically rotating fluids. II. J. Math. Phys. **18**, 1206–1209 (1977)

Harris, E.G.: Analogy between general relativity and electromagnetism for slowly moving particles in weak gravitational fields. Am. J. Phys. **59**, 421–425 (1991)

Hawking, S.W.: On the rotation of the Universe. Observatory **89**, 38–39 (1969)

Hennig, J., Neugebauer, G.: Non-existence of stationary two-black-hole configurations: the degenerate case. Gen. Relativ. Gravit. **43**, 3139–3162 (2011)

Hofmann, K.-D.: Über Wechselwirkungen von Gravitation und elektromagnetischem Feld gemäß der allgemeinen Relativitätstheorie. Z. Phys. **166**, 567–576 (1962)

Hönl, H., Maue, A.: Über das Gravitationsfeld rotierender Massen. Z. Phys. **144**, 152–167 (1956)

Katz, J., Lynden-Bell, D., Bičák, J.: Instantaneous inertial frame but retarded electromagnetism in rotating relativistic collapse. Class. Quantum Gravit. **15**, 3177–3194 (1998)

King, D.H.: A closed universe cannot rotate. In: Barbour, J., Pfister, H. (eds.) Mach's Principle—From Newton's Bucket to Quantum Gravity, pp. 237–246. Birkhäuser, Boston (1995)

King, M., Pfister, H.: Electromagnetic Thirring problems. Phys. Rev. D **63**, 104004 (2001)

Klein, C.: Rotational perturbations and frame dragging in a Friedmann universe. Class. Quantum Gravit. **10**, 1619–1631 (1993)

Klein, C.: Exact relativistic treatment of stationary counterrotating dust disks: boundary value problems and solutions. Phys. Rev. D **63**, 064033 (2001)

Klein, M.J., et al. (eds.): The Collected Papers of Albert Einstein, vol. 5. Princeton University Press, Princeton (1993)

Klein, M.J., et al. (eds.): The Collected Papers of Albert Einstein, vol. 4. Princeton University Press, Princeton (1995)

Kogut, A., Hinshaw, G., Banday, A.J.: Limits to global rotation and shear from the COBE DMR four-year sky maps. Phys. Rev. D **55**, 1901–1905 (1997)

Kovalevsky, J., Mueller, I.I., Kolaczek, B. (eds.): Reference Frames in Astronomy and Geophysics. Kluwer Academic, Dordrecht (1989)

Kovalevsky, J., et al.: The Hipparcos catalogue as a realization of the extragalactic reference frame. Astron. Astrophys. **323**, 620–633 (1997)

Lanczos, K.: Zum Rotationsproblem der allgemeinen Relativitätstheorie. Z. Phys. **14**, 204–219 (1923)

Lense, J., Thirring, H.: Über den Einfluß der Eigenrotation der Zentralkörper auf die Bewegung der Planeten und Monde nach der Einsteinschen Gravitationstheorie. Phys. Z. **19**, 156–163 (1918). English translation in Gen. Relativ. Gravit. **16**, 727–741 (1984)

Lewis, S.M.: Machian effects in nonasymptotically flat space-times. Gen. Relativ. Gravit. **12**, 917–924 (1980)

Lindblom, L.A.: Fundamental properties of equilibrium stellar models. Ph.D. thesis, University of Maryland (1978)

Lindblom, L., Brill, D.R.: Inertial effects in the gravitational collapse of a rotating shell. Phys. Rev. D **10**, 3151–3155 (1974)

Lynden-Bell, D., Katz, J.: Thought experiments on gravitational forces. Mon. Not. R. Astron. Soc. **438**, 3163–3176 (2014)

Lynden-Bell, D., Katz, J., Bičák, J.: Mach's principle from the relativistic constraint equations. Mon. Not. R. Astron. Soc. **272**, 150–160 (1995)

Lynden-Bell, D., Bičák, J., Katz, J.: On accelerated inertial frames in gravity and electromagnetism Ann. Phys. **271**, 1–22 (1999)

Mach, E.: Die Geschichte und die Wurzel des Satzes von der Erhaltung der Arbeit. Calve, Prag (1872)

Mach, E.: Die Mechanik in ihrer Entwicklung. Historisch-kritisch dargestellt. F. A. Brockhaus, Leipzig (1883)

Matte, A.: Sur de nouvelles solutions oscillatoires des équations de la gravitation. Canad. J. Math. **5**, 1–16 (1953)

Meinel, R., Kleinwächter, A.: Dragging effects near a rigidly rotating disc of dust. In: Barbour, J.B., Pfister, H. (eds.) Mach' Principle—From Newton's Bucket to Quantum Gravity. Birkhäuser, Boston, 339–346 (1995)

Misner, C.W., Thorne, K.S., Wheeler, J.A.: Gravitation. Freeman, San Francisco (1973)

Neugebauer, G., Hennig, J.: Non-existence of stationary two-black-hole configurations. Gen. Relativ. Gravit. **41**, 2113–2130 (2009)

Neugebauer, G., Meinel, R.: General relativistic gravitational field of a rigidly rotating disk of dust: axis potential, disk metric, and surface mass density. Phys. Rev. Lett. **73**, 2166–2168 (1994)

Neumann, C.: Über die Principien der Galilei-Newtonschen Theorien. Teubner, Leipzig (1870)

Ohanian, H.C., Ruffini, R.: Gravitation and Spacetime, 3rd edn. Cambridge University Press, Cambridge (2013)

Overduin, J.: The experimental verdict on spacetime from gravity probe B. In: Petkov, V. (ed.) Space, Time, and Spacetime, Fundamental Theories of Physics, vol. 167, pp. 25–59. Springer, Berlin (2010)

Ozsváth, I., Schücking, E.: An anti-Mach metric. In: Recent Developments in General Relativity, pp. 339–350. Pergamon Press, Oxford (1962)

Pais, A.: 'Subtle is the Lord ...' The Science and the Life of Albert Einstein. Oxford University Press, Oxford (1982)

Pareja, M.J.: Relativistic stars in differential rotation: bounds on the dragging rate and on the rotational energy. J. Math. Phys. **45**, 677–695 (2004a)

Pareja, M.J.: Bounds on the dragging rate of slowly and differentially rotating relativistic stars. J. Math. Phys. **45**, 3379–3398 (2004b)

Petrov, A.Z.: Einstein Spaces. Pergamon Press, Oxford (1969)

Pfister, H.: Rotating mass shells with flat interiors. Class. Quantum Gravit. **6**, 487–504 (1989)

Pfister, H.: Newton's first law revisited. Found. Phys. Lett. **17**, 49–64 (2004)

Pfister, H.: On the history of the so-called Lense–Thirring effect. Gen. Relativ. Gravit. **39**, 1735–1748 (2007)

Pfister, H.: The history of the so-called Lense–Thirring effect, and of related effects. In: Ciufolini, I., Matzner, R.A. (eds.) General Relativity and John Archibald Wheeler, pp. 493–503. Springer, Berlin (2010)

Pfister, H.: Editorial note to: Hans Thirring, On the formal analogy between the basic electromagnetic equations and Einstein's gravity equations in first approximation. Gen. Relativ. Gravit. **44**, 3217–3224 (2012)

Pfister, H.: Gravitomagnetism: from Einstein's 1912 paper to the satellites LAGEOS and gravity probe B. In: Bičák, J., Ledvinka, T. (eds.) Relativity and Gravitation—100 Years After Einstein in Prague. Springer Proceedings in Physics, vol. 157, pp. 191–197. Springer, Berlin (2014)

Pfister, H., Braun, K.H.: Induction of correct centrifugal force in a rotating mass shell. Class. Quantum Gravit. **2**, 909–918 (1985)

Pfister, H., Braun, K.H.: A mass shell with flat interior cannot rotate rigidly. Class. Quantum Gravit. **3**, 335–345 (1986)

Pfister, H., King, M.: Rotating charged mass shell: dragging, antidragging, and the gyromagnetic ratio. Phys. Rev. D **65**, 084033 (2002)

Pfister, H., Schedel, Ch.: Can gravitational attraction be compensated for by gravimagnetic effects?. Class. Quantum Gravit. **4**, 141–147 (1987)

Pfister, H., Frauendiener, J., Hengge, S.: A model for linear dragging. Class. Quantum Gravit. **22**, 4743–4761 (2005)

Rindler, W.: Essential Relativity. Van Nostrand-Reinhold, New York (1969)

Schiff, L.I.: Possible new experimental test of general relativity theory. Phys. Rev. Lett. **4**, 215–217 (1960)

Schmid, C.: Cosmological Vorticity Perturbations, Gravitomagnetism, and Mach's Principle. E-print arXiv: 0201095 [gr-qc] (2002)

Schmid, C.: Cosmological gravitomagnetism and Mach's principle. Phys. Rev. D **74**, 044031 (2006)

Schmid, C.: Mach's principle: exact frame-dragging via gravitomagnetism in perturbed Friedmann–Robertson–Walker universes with $K = (\pm 1, 0)$. Phys. Rev. D **79**, 064007 (2009)

Schreiber, K.U., Wells, J.-P.R., Stedman, G.E.: Noise processes in large ring lasers. Gen. Relativ. Gravit. **40**, 935–943 (2008)

Schücking, E.L.: Gravitation and inertia. Phys. Today, June 1996, 58

Schulmann, R., et al. (eds.): The Collected Papers of Albert Einstein, vol. 8. Princeton University Press, Princeton (1998)

Seeliger, H.: Über die sogenannte absolute Bewegung. Sitzb. Königl. Bayer. Akad. d. Wiss. math.-phys. Kl. **36**, 85–137 (1906)

Shapiro, I.I., Reasenberg, R.D., Chandler, J.F., Babcock, R.W.: Measurement of the de sitter precession of the moon: a relativistic three-body effect. Phys. Rev. Lett. **61**, 2643–2646 (1988)

Soergel-Fabricius, C.: Über den Ursprung von Coriolis- und Zentrifugalkräften in stationären Räumen. Z. Phys. **161**, 392–403 (1961)

Stedman, G.E.: Ring-laser tests of fundamental physics and geophysics. Rep. Prog. Phys. **60**, 615–688 (1997)

Stella, L., Possenti, A.: Lense–Thirring precession in the astrophysical context. Space Sci. Rev. **148**, 105–121 (2009)

Thirring, H.: Wirkung Rotierender Massen. Notebook. Österr. Zentralbibliothek, Wien (1917)

Thirring, H.: Über die Wirkung rotierender ferner Massen in der Einsteinschen Gravitationstheorie. Phys. Zs. **19**, 33–39 (1918). English translation in Gen. Relativ. Gravit. **16**, 712–725 (1984)

Thirring, H.: Über die formale Analogie zwischen den elektromagnetischen Grundgleichungen und den Einsteinschen Gravitationsgleichungen erster Näherung. Phys. Z. **19**, 204–205 (1918). English translation In: Gen. Relativ. Gravit. **44**, 3225–3229 (2012)

Thirring, H.: Berichtigung zu meiner Arbeit: "Über die Wirkung rotierender Massen in der Einsteinschen Gravitationstheorie". Phys. Z. **22**, 29–30 (1921)

van Zoest, T., et al.: Bose–Einstein condensation in microgravity. Science **328**, 1540–1543 (2010)

Weinberg, S.: Gravitation and Cosmology: Principles and Applications of the General Theory of Relativity. Wiley, New York (1972)

Wheeler, J.A.: Geometrodynamics and the issue of the final state. In: deWitt, C., deWitt, B. (eds.) Relativity, Groups and Topology. Gordon and Breach, New York (1964)

Appendix A
A Sketch of the Proof that the Inertial Path Structure Follows from a Local Desargues Property

In this appendix we present a sketch, and important details, of the proof that a path structure which obeys the Desargues incidence properties up to order ϵ^2 in a two-dimensional ϵ-neighborhood of a 'central point' e in spacetime is a free fall structure, i.e., it is linear up to order ϵ^2. More details can be found in Heilig and Pfister (1990, Sect. 4).

If, in agreement with EPS, we assume that the free particle paths are one-dimensional C^3-submanifolds of $\mathcal{M}$, then such a path through an event p (in the ϵ-neighborhood of e, which we choose as the origin of the two-dimensional coordinate system) can be Taylor expanded in the path parameter τ :

$$x^\mu(\tau) = x^\mu(0) + \tau v^\mu(0) + \tau^2 a^\mu(x^\nu, v^\lambda) + O(\epsilon^3)$$
$$\approx x^\mu(0) + \tau v^\mu(0) + \tau^2 a^\mu(0, v^\nu) + O(\epsilon^3) , \tag{A.1}$$

where $K^\mu(v^\nu) := a^\mu(0, v^\nu)$ is the 'acceleration field'. Because the path depends only on the direction at p, and not on the 'length' of $v^\mu(0)$, we have the scaling law $K^\mu(\lambda v^\nu) = \lambda^2 K^\mu(v^\nu)$ for all $\lambda \neq 0$. With $\tau = \epsilon\rho$ and ρ finite (non-infinitesimal, but such that the path does not leave the ϵ-neighborhood of the origin), (A.1) takes the form

$$x^\mu(\epsilon\rho) = x^\mu(0) + \epsilon\rho v^\mu + \epsilon^2\rho^2 K^\mu(v^\nu) + O(\epsilon^3) , \tag{A.2}$$

and if this path connects (exactly to order ϵ^2) the events p and q, we have

$$x^\mu(\epsilon\rho) = x_p^\mu + \rho v^\mu + \epsilon^2\rho(\rho - 1)K^\mu(v^\nu) + O(\epsilon^3) , \tag{A.3}$$

with $v^\mu = x_q^\mu - x_p^\mu$. In a two-dimensional coordinate system, the decisive last term of this equation can be considerably simplified. By an appropriate parameter transformation $\rho \rightarrow \sigma$, we can achieve that the component K^2 vanishes, and

© Springer International Publishing Switzerland 2015
H. Pfister, M. King, *Inertia and Gravitation*, Lecture Notes in Physics 897,
DOI 10.1007/978-3-319-15036-9

a rescaling of $K^\mu = (K^1, K^2)$ can have the result that K^1 depends only on v^2/v^1:

$$x^\mu(\epsilon\sigma) = x_p^\mu + \sigma v^\mu + \sigma(\sigma - 1)(v^1)^2 \begin{pmatrix} g(v^2/v^1) \\ 0 \end{pmatrix} + O(\epsilon^3). \tag{A.4}$$

It turns out that, to prove the 'if' part of Axiom P_2 in Sect. 2.3 in two dimensions, we do not need the most general Desargues configuration. It suffices to consider the case where the path P is the x^1-axis, P' the x^2-axis, P'' the path defined by e and A'', and the points $A, \ldots, B''$ have the coordinates $A = \epsilon(1, 0)$, $B = \epsilon(-1, 0)$, $A' = \epsilon(0, a)$, $B' = \epsilon(0, 1)$, $A'' = \epsilon(1, 1)$, and $B'' = \epsilon(c, c) + \epsilon^2 c(c - 1)(g(1), 0)$, where a and c are largely arbitrary real numbers, but such that the whole Desargues figure is confined to an ϵ-neighborhood of the origin. The events e_1, e_2, e_3 are then easily found (up to errors of order ϵ^3) as the intersection points of pairs of paths of the form (A.4), according to Fig. 1.2 of Sect. 1.4:

$$e_1 = \frac{\epsilon}{1 + a} \begin{pmatrix} a - 1 \\ 2a \end{pmatrix} + \frac{2a(1 - a)\epsilon^2}{(1 + a)^3} \begin{pmatrix} g(-a) - g(1) \\ g(-a) + ag(1) \end{pmatrix},$$

$$e_2 = \frac{\epsilon}{1 + c} \begin{pmatrix} 1 + c \\ 2c \end{pmatrix} + \frac{2c(1 - c)\epsilon^2}{(1 + c)^3} \begin{pmatrix} 0 \\ c(1 + c)g(1) - (1 + c)^2 g(c/(1 + c)) \end{pmatrix},$$

$$e_3 = \frac{\epsilon}{1 - ac} \begin{pmatrix} c(1 - a) \\ a + c - 2ac \end{pmatrix}$$

$$+ \frac{c(1 - a)^2(c - 1)^2\epsilon^2}{(1 - ac)^3} \begin{pmatrix} \frac{c}{c - 1}g(1 - a) - \frac{ac}{1 - a}g(\frac{c - 1}{c}) - \frac{1 - ac}{1 - a}g(1) \\ g(1 - a) - acg(\frac{c - 1}{c}) - (1 - ac)g(1) \end{pmatrix}.$$

Now, the Desargues theorem requires that e_3 lies, up to order ϵ^2, on the path defined by e_1 and e_2. This leads to the following functional equation for the function $g(x)$:

$$c(1 + 2a)(1 - ac)g(1) = c(1 - a^2)g(1 - a) + ac^2(c - 1)(1 + a)g\left(\frac{c - 1}{c}\right)$$

$$-a(1 - ac)g(-a) + (1 - ac)(1 + a)c(1 + c)g\left(\frac{c}{1 + c}\right)$$

$$-(c - a)(1 + c)g\left(\frac{c - a}{1 + c}\right). \tag{A.5}$$

In the review Rund et al. (1992) of the paper Heilig and Pfister (1990), the well-known mathematician H. Rund has called this a 'formidable functional equation'. But he has also praised the paper as "a most ingenious application of the concepts of projective geometry to a study of the foundations of classical mechanics, special relativity, and the general theory of relativity".

It is easily checked that $g(x) \equiv g(1) =$ const. satisfies (A.5). This leads to a slightly simpler functional equation for the function $f(x) = g(1 - x) - g(1)$:

$$0 = c(1 - a^2)f(a) + ac^2(1 + a)(c - 1)f\left(\frac{1}{c}\right) - a(1 - ac)f(1 + a)$$

$$-(c-a)(1 + c)f\left(\frac{1 + a}{1 + c}\right) + (1-ac)(1 + a)c(1 + c)f\left(\frac{1}{1 + c}\right). \quad (A.6)$$

A somewhat technical proof in Heilig and Pfister (1990) leads to the relation $f(x) = x^2 f(1/x)$, and this together with (A.6) results finally in $f(x) = \alpha x$, with a constant α, or in $g(x) = \beta x - \gamma$, with constants β, γ. Inserting this into (A.4) and performing appropriate coordinate and parameter transformations nullifies the acceleration term of this equation and proves that all paths in the two-dimensional Desargues configuration are linear paths up to order ϵ^2. The extension of the proof to higher dimensions is, as mentioned, quite formal, due to the surface-forming property of pairs of paths like AA'' and AA'' in Fig. 1.2 of Sect. 1.4. The coordinates of the events $A, \ldots, B''$ must of course be chosen anew (but again in a special way), and appropriate coordinate and parameter transformations must then be applied once again. But no new functional equation like (A.5) or (A.6) has to be solved. (For details see Heilig and Pfister 1990, Sect. 4.)

References

Heilig, U., Pfister, H.: Characterization of free fall paths by a global or local Desargues property. J. Geom. Phys. **7**, 419–446 (1990)

Rund, H., Heilig, U., Pfister, H.: Characterization of free fall paths by a local or global Desargues property. Math. Rev. **92h**, 83008 (1992)

Appendix B
Slowly Rotating Mass Shells with Flat Interiors

In the papers (Pfister and Braun 1985, 1986) stationary and axisymmetric rotating mass shells with flat interiors, with possibly high mass values M, radius R, and small angular velocities ω have already been calculated mathematically (as power series in $\omega R \ll 1$), and physically analyzed, in particular concerning Mach's idea of relativity of rotation. Here we repeat and summarize some of these results, but we no longer calculate the energy-momentum tensor of the shell material using the discontinuities of the derivatives of the metric functions at the shell position but by using the more elegant, mainly geometric formalism of Israel (1966) which expresses the energy-momentum tensor of the shell in terms of the extrinsic curvatures of the embeddings of the mass shell Σ in the different vacuum spacetimes V^+ and V^- for the exterior and interior of the shell. We also correct some errors in the papers (Pfister and Braun 1985, 1986). Although in the Israel-formalism it is not necessary to have a continuous metric across the shell, we nevertheless use the metric form of Pfister and Braun (1985):

$$ds^2 = g_{\mu\nu}dx^\mu dx^\nu = -e^{2U}dt^2 + e^{-2U}[e^{2K}(dr^2 + r^2 d\vartheta^2) + W^2(d\varphi - \omega\, A dt)^2],$$
$$\text{(B.1)}$$

where the metric functions U, K, W, A are functions of r and ϑ only, due to stationarity and axial symmetry of our models. This form provides a continuous metric across the shell, and we can use the explicit solutions of Einstein's vacuum field equations given in Pfister and Braun (1985, 1986). Furthermore, it turns out that starting (for $\omega = 0$) with the Schwarzschild mass shell in Schwarzschild coordinates, besides being discontinuous, would lead to algebraically more complicated expressions for the energy-momentum tensor and for the metric corrections in higher orders in ω, containing square roots [compare Pfister and Braun (1985) with, e.g., Pfister et al. (2005)].

© Springer International Publishing Switzerland 2015
H. Pfister, M. King, *Inertia and Gravitation*, Lecture Notes in Physics 897,
DOI 10.1007/978-3-319-15036-9

A complete set of independent vacuum field equations for the metric (B.1) reads

$$\Delta W = 0, \tag{B.2}$$

$$W \Delta U + (U_1 W_1 + \tfrac{1}{r^2} U_2 W_2) = \tfrac{\omega^2}{2} W^3 e^{-4U} (A_1^2 + \tfrac{1}{r^2} A_2^2), \tag{B.3}$$

$$W_{12} - \tfrac{1}{r} W_2 + 2W U_1 U_2 - K_1 W_2 - K_2 W_1 = \tfrac{\omega^2}{2} W^3 e^{-4U} A_1 A_2, \tag{B.4}$$

$$W_{11} + W(U_1^2 - \tfrac{1}{r^2} U_2^2) - K_1 W_1 + \tfrac{1}{r^2} K_2 W_2 = \tfrac{\omega^2}{4} W^3 e^{-4U} (A_1^2 - \tfrac{1}{r^2} A_2^2), \tag{B.5}$$

$$\Delta A + \tfrac{3}{W}(A_1 W_1 + \tfrac{1}{r^2} A_2 W_2) - 4(U_1 A_1 + \tfrac{1}{r^2} U_2 A_2) = 0, \tag{B.6}$$

where a subscript 1 denotes the r-derivative and a subscript 2 the ϑ-derivative, and where $\Delta = \partial^2/\partial r^2 + r^{-1}\partial/\partial r + r^{-2}\partial^2/\partial\vartheta^2$ is the flat Laplacian. Obviously, (B.2)–(B.5) are relevant for the even orders in ω in our power series expansion, and (B.6) for the odd orders.

With the dimensionless variables $x = 2r/M$ and $X = 2R/M$, it is well known from the literature that, in the static limit of the Schwarzschild shell, we have (in isotropic coordinates):

$$\overset{0}{U}(x) = \log \frac{x-1}{x+1} \text{ for } V^+(x > X), \qquad \overset{0}{U} = \log \frac{X-1}{X+1} \text{ for } V^-(x < X),$$
$$\tag{B.7}$$

$$\overset{0}{K}(x) = \log \frac{x^2-1}{x^2} \text{ for } V^+(x > X), \qquad \overset{0}{K} = \log \frac{X^2-1}{X^2} \text{ for } V^-(x < X),$$
$$\tag{B.8}$$

$$\overset{0}{W}(x,\vartheta) = e^{\overset{0}{K}(x)} r \sin\vartheta. \tag{B.9}$$

As basis vectors in $\Sigma(r = R)$ we choose $e_t^\mu = (1,0,0,0)$, $e_\vartheta^\mu = (0,0,1,0)$, and $e_\varphi^\mu = (0,0,0,1)$. The unit normal vector to Σ is $n_\mu = (0, (X+1)^2/X^2, 0, 0)$. [In Schwarzschild coordinates the time coordinate t, the basis vector e_t^μ and n_μ would be discontinuous at $r = R$ (Pfister et al. 2005)!] According to Israel (1966) the symmetric extrinsic curvature 3-tensors in V^+ and V^- are given by

$$K_{ab} = n_\mu e_{a;b}^\mu, \tag{B.10}$$

with $(a,b) \in (t,\vartheta,\varphi)$. Since in the static case n_μ has only an r-component, and the vectors e_a^μ are constant, we have $\overset{0}{K}_{ab} = n_r \, {}^{(4)}\Gamma_{bc}^r e_a^c$, and since only diagonal metric-components are present, and only the r-derivatives count, the evaluation is relatively simple, with the result:

$$R \overset{0}{K}{}^{+}_{tt} = \frac{2X^3(X-1)}{(X+1)^5}, \quad \frac{\overset{0}{K}{}^{+}_{\vartheta\vartheta}}{R} = \frac{\overset{0}{K}{}^{+}_{\varphi\varphi}}{R\sin^2\vartheta} = -\frac{X^2-1}{X^2},$$

$$\frac{\overset{0}{K}{}^{-}_{\vartheta\vartheta}}{R} = \frac{\overset{0}{K}{}^{-}_{\varphi\varphi}}{R\sin^2\vartheta} = -\frac{(X+1)^2}{X^2}. \tag{B.11}$$

All other $\overset{0}{K}_{ab}$ are zero in the static case. According to Israel (1966) one defines $\gamma_{ab} = K^+_{ab} - K^-_{ab}$, whence the energy-momentum tensor becomes $8\pi\,S^a_b = \gamma^a_b - \delta^a_b\,\gamma^c_c$, with the explicit results

$$8\pi\,R\overset{0}{S}{}^{t}_{t} = -\frac{4X^2}{(X+1)^3}, \quad 8\pi\,R\overset{0}{S}{}^{\vartheta}_{\vartheta} = 8\pi\,R\overset{0}{S}{}^{\varphi}_{\varphi} = \frac{2X^2}{(X+1)^3(X-1)}. \tag{B.12}$$

These terms coincide with the results of (2.12) in Pfister and Braun (1985), if one takes into consideration the fact that there the components of the energy-momentum tensor were multiplied by a function $\delta(r-R)$.

According to Pfister and Braun (1985), to fist order in ω, the metric function $\overset{0}{A}(r,\vartheta)$ is independent of ϑ. To solve (B.6), it is advantageous to introduce the variables $y = 4x/(x+1)^2$ and $Y = 4X/(X+1)^2$. These variables are also preferable for some of the later analysis of the higher orders in ω, but not in all expressions, because otherwise square roots $\sqrt{1-y}$ would appear. Then to order ω, (B.6) takes the form

$$\left(\frac{d^2}{dy^2} - \frac{2}{y}\frac{d}{dy}\right)\overset{0}{A}(y) = 0. \tag{B.13}$$

From the two solutions we have to choose the one which falls off asymptotically, because we define rotation relative to the asymptotic rest frame,

$$\overset{0}{A}(y) = \lambda\left(\frac{y}{4}\right)^3 \text{ for } V^+(y > Y), \quad \overset{0}{A} = \lambda\left(\frac{Y}{4}\right)^3 \text{ for } V^-(y < Y), \tag{B.14}$$

with a constant λ. [This constant and the later integration constants are defined differently than in Pfister and Braun (1985, 1986)!] For the extrinsic curvature components $\overset{1}{K}_{t\varphi}$ we then get

$$\overset{1}{K}{}^{+}_{t\varphi} = -\omega R\lambda\frac{X(X-1)}{2(X+1)^5}\sin^2\vartheta, \quad \overset{1}{K}{}^{-}_{t\varphi} = \omega R\lambda\frac{X}{(X+1)^4}\sin^2\vartheta, \tag{B.15}$$

and

$$8\pi \overset{1}{S}{}^{\varphi}_{t} = -\frac{\omega\lambda X^{5}(3X-1)}{2R(X+1)^{8}(X-1)}. \tag{B.16}$$

The constant λ must of course to be related to the angular velocity ω in such a way that the energy current S^{φ}_{t} describes the rigid and axial rotation of the spherical mass shell. (A geometric deformation and a nonspherical mass distribution of the shell happen only in the higher even orders of ω. See below.) This means mathematically that the four-velocity of the matter has the form $u^{\mu} = u^{t}(1,0,0,\omega)$, and is an eigenvector of the energy-momentum tensor: $S^{\mu}_{\nu}u^{\nu} = -\rho u^{\mu}$, with the invariant mass density $\rho = -S^{t}_{t}$. From the equation for $\mu = \varphi$ we get

$$\lambda = 4\frac{(X+1)^{5}(2X-1)}{X^{3}(3X-1)}, \tag{B.17}$$

being equivalent to (2.15) in Pfister and Braun (1985).

To order ω^{2} the solution $\overset{0}{A}(r)$ from (B.14) operates as a source term in the field equations (B.2)–(B.5). It is therefore clear that in the expansions

$$N(r,\vartheta) = \overset{0}{N}(r,\vartheta) + (\omega R)^{2}\overset{2}{N}(r,\vartheta) + (\omega R)^{4}\overset{4}{N}(r,\vartheta) + \ldots$$

for $N \in (W,U,K,A)$, expanded in addition in Legendre polynomials $P_{l}(\cos\vartheta)$, the potentials $\overset{2}{N}(r,\vartheta)$ contain only P_{0} and P_{2}, similar to Hartle's work (Hartle 1967) on slowly rotating stars. [The uneven Legendre polynomials are missing due to the equatorial symmetry of the problem. These expansions in powers of ωR, i.e., of the velocity of the shell matter in units of the light velocity, seem to be more appropriate than the expansions in powers of ωM in Pfister and Braun (1986).] The functions $\overset{2}{N}(r,\vartheta)$, $\overset{4}{N}(r,\vartheta)$, and so on, must of course fall off asymptotically, in order to realize Minkowski geometry there. They must even fall off at least as fast as the functions $\overset{0}{N}(r)$, in order that the power series expansion in ω should also be consistent asymptotically, i.e., $\overset{2}{U}$, $\overset{4}{U}$, and so on, must fall off at least like r^{-1}, then $\overset{2}{K}$, $\overset{4}{K}$, and so on, at least like r^{-2}, and $\overset{2}{A}$, $\overset{4}{A}$, and so on, at least like r^{-3}. We even demand that $\overset{2}{U}$, $\overset{4}{U}$, and so on, should fall off more quickly than r^{-1}, in order that the mass value M is fixed by the function $\overset{0}{U}(r)$, and is not corrected by higher orders in ω. According to (B.2), the function $W_{,22}(\vartheta)$ has to reproduce the ϑ-behavior of $W(\vartheta)$. A complete set of such functions is the series $\{\sin n\vartheta, \cos n\vartheta\}$ with $n = 1,2,3,\ldots$. The functions $\cos n\vartheta$ drop out because they would lead to a singularity of the metric (B.1) at the poles $\vartheta = 0,\pi$. For the functions $\sin n\vartheta$ only odd n are allowed due to the equatorial symmetry. For a function $\sin n\vartheta$ the x-behavior can only be x^{n} and x^{-n}, according to (B.2). However, the positive powers are forbidden due to the asymptotic behavior, and because the total mass M should not attain corrections from higher orders of ω. Therefore, the only consistent solution of (B.2) to order ω^2 is

$$\overset{2}{W}(x, \vartheta) = M\left(\frac{\beta_0}{x} \sin\vartheta + \frac{\beta_2}{x^3} \sin 3\vartheta\right) = M \sin\vartheta \left[\frac{\beta_0}{x} + \frac{\beta_2}{x^3}(3 - 4\sin^2\vartheta)\right],$$

(B.18)

with integration constants β_0, β_2.

With $\overset{2}{U}(x, \vartheta) = g(x) + h(x)P_2(\cos\vartheta)$, (B.3) splits into the two differential equations:

$$h'' + \frac{2x}{x^2 - 1}h' - \frac{6}{x^2}h = \frac{64\beta_2(2x^2 - 1)}{3x^3(x^2 - 1)^3} - \frac{3\lambda^2 x^2}{X^2(x + 1)^8},$$

(B.19)

$$g'' + \frac{2x}{x^2 - 1}g' = \frac{8\beta_0 x}{(x^2 - 1)^3} + \frac{8\beta_2(2x^2 - 1)}{3x^3(x^2 - 1)^3} + \frac{3\lambda^2 x^2}{X^2(x + 1)^8},$$

(B.20)

where h', h'', etc. denote the derivatives of these functions with respect to x. The asymptotically decreasing solutions are:

$$h(x) = \frac{16\beta_2}{3x(x^2 - 1)} + \frac{\lambda^2 x^3}{12X^2(x + 1)^6}$$

$$+ \gamma_2\left[\frac{2(x^2 + 1)}{x} + \left(x^2 + \frac{2}{3} + \frac{1}{x^2}\right)\log\frac{x - 1}{x + 1}\right],$$

(B.21)

$$g(x) = \frac{2\beta_0 x}{x^2 - 1} + \frac{2\beta_2(3x^2 - 2)}{3x(x^2 - 1)} - \frac{\lambda^2 x(x^4 + 6x^3 + 46x^2/3 + 6x + 1)}{64X^2(x + 1)^6}$$

$$+ \gamma_0 \log\frac{x - 1}{x + 1},$$

(B.22)

with integration constants γ_2 and γ_0. To ensure that the total mass M of the shell is not changed in order ω^2, the term $\overset{2}{U}(x, \vartheta)$ has to fall off asymptotically faster than x^{-1}, which requires $\gamma_0 = \beta_0 + \beta_2 - \lambda^2/128X^2$.

In order to find the solution $\overset{2}{K}(x, \vartheta)$, we first consider (B.4) to order ω^2. After inserting $\overset{2}{W}$ and $\overset{2}{U}$, and taking into account the 'angular momentum cutoff' in the form $\overset{2}{K}(x,\vartheta) = k(x) + l(x)\sin^2\vartheta$, (B.4) breaks down into the two first order differential equations:

$$k'(x) = -\frac{4x}{(x^2-1)^2}\left[\beta_0 + \frac{3\beta_2}{x^4}(2x^2-1)\right], \tag{B.23}$$

$$(x^2-1)l'(x) + \frac{2}{x}(x^2+1)l(x) = \frac{48\beta_2(2x^2-1)}{x^3(x^2-1)} - 12h(x). \tag{B.24}$$

The asymptotically decreasing solutions are:

$$k(x) = \frac{2\beta_0}{x^2-1} + \frac{6\beta_2}{x^2(x^2-1)}, \tag{B.25}$$

$$l(x) = -\frac{4\beta_2(4x^2-3)}{x^2(x^2-1)^2} + \frac{\lambda^2 x^4}{2X^2(x+1)^6(x-1)^2} -$$

$$- 4\gamma_2\left[2\frac{x^4+1}{(x^2-1)^2} + \frac{x^2+1}{x}\log\frac{x-1}{x+1}\right] + \delta_2\frac{x^2}{(x^2-1)^2}, \tag{B.26}$$

with a new integration constant δ_2. Equation (B.5) provides the following relations between the integration constants:

$$\gamma_2 = -\frac{3}{8}\gamma_0, \qquad \delta_2 = -2\gamma_0 - 4\beta_0. \tag{B.27}$$

Having solved all relevant field equations in the vacuum exterior of the rotating mass shell to order ω^2, we have to perform the (continuous!) connection to a flat interior metric of the form

$$ds^2 = -e^{2\hat{U}}dt^2 + e^{2\hat{K}-2\hat{U}}[dr^2 + r^2 d\vartheta^2 + r^2\sin^2\vartheta(d\varphi - \omega\hat{A}dt)^2], \tag{B.28}$$

with constants

$$\hat{U} = \overset{0}{U}(X) + (\omega R)^2\overset{2}{U}_0, \quad \hat{K} = \overset{0}{K}(X) + (\omega R)^2\overset{2}{K}_0, \quad \hat{A} = \overset{0}{A}(X) + (\omega R)^2\overset{2}{A}_0.$$

If this should happen at a spherical shell shape $r = R$, we would have to fulfil the relation $\overset{2}{W}(X) = (M(X^2-1)/2X)\overset{2}{K}(X)\sin\vartheta$. From the continuity of $\overset{2}{U}$ and $\overset{2}{K}$ at $x = X$ (independent of ϑ!) it follows that $h(X) = 0$ and $l(X) = 0$. From the above expression for $\overset{2}{W}(X)$ it then also follows that $\beta_2 = 0$. Inserting these results into (B.21) and (B.26) for $h(X)$ and $l(X)$ leads to two expressions for γ_2

which contradict each other. A consistent solution is possible only if we allow for a non-spherical shape of the shell of the form

$$r_S = R[1 + (\omega R)^2 f \sin^2 \vartheta]. \tag{B.29}$$

Here f is some constant or some function of the physical parameters R and M of our model. And since f is dimensionless, it can only depend on the dimensionless quantity $X = 2R/M$. Such a non-spherical (centrifugal) deformation of the shell is to be expected on physical grounds anyway. Besides (B.25) for $x = X$, the continuity conditions for $U(r, \vartheta)$, $K(r, \vartheta)$ and $W(r, \vartheta)$ at the shell position r_S lead on the one hand to the conditions

$$\overset{2}{U}_0 = g(X) + h(X), \qquad \overset{2}{K}_0 = k(X). \tag{B.30}$$

On the other hand, and more interestingly, from the ϑ-dependent terms we get

$$h(X) = \frac{4X}{3(X^2 - 1)} f(X), \, l(X) = -\frac{2}{X^2 - 1} f(X),$$

$$\beta_2 = \frac{X^2}{4} f(X), \, l'(X) = \frac{4(3X^2 - 2)}{X(X^2 - 1)^2} f(X). \tag{B.31}$$

Inserting the last expressions into (B.21) and (B.26), observing the values λ and δ_2 from (B.17) and (B.27), and eliminating the constant γ_2, leads finally to the unique, although somewhat involved result for the asphericity parameter

$$f(X) = \frac{(X + 1)^4 (2X - 1)^2}{2X^4 (3X - 1)^2}$$

$$\left\{ (1 + \frac{6}{X} + \frac{1}{X^2}) - \frac{32 \left[2X + (X^2 + 1) \log \frac{X-1}{X+1} \right]}{3 \left[2X(X^2 + 1) + (X^4 + \frac{2}{3}X^2 + 1) \log \frac{X-1}{X+1} \right]} \right\} \tag{B.32}$$

A graph of this parameter in the physical region $X \geq 1$ was given in Pfister and Braun (1985). Here we list only its most important physical properties:

(a) f is negative for all $X > 1$, so that the coordinate radius r of the shell is smaller in the equatorial plane than in the polar directions. Also the invariant equatorial circumference is smaller than a polar circumference by an amount $\pi R(\omega R)^2 ((X+1)(X+3)/X^2)(-f(X))$. Near $X = 1.25$ this ellipticity reaches its maximum. [Qualitatively, these results coincide with Pfister and Braun (1985), but quantitatively there are differences to the—obviously wrong— results in Pfister and Braun (1985).]

(b) In the weak-field limit $X \to \infty$ the parameter f reaches the non-zero value -2, in accordance with the fact that H. Thirring (Thirring 1918) and others could not get a flat interior of the shell, and therefore no correct centrifugal force inside their spherical mass shell.

(c) For $X \to 1$, i.e., for a collapsing shell f goes to zero. This agrees with a result in de la Cruz and Israel (1968), who investigated (up to order ω^3) whether rotating mass shells can be the source of the Kerr metric. They obtained the result that sphericity and rigid rotation of the shell, and flat inside geometry can only be reached in the limit $X \to 1$. In this limit, our constants have the values $\beta_0 = 32$, $\beta_2 = \gamma_0 = \gamma_2 = 0$, $\delta_2 = -128$, and the resulting metric coincides with the Kerr metric up to order ω^2. However, for $X > 1$, our metric differs from the Kerr metric, what should be no surprise, since the Kerr metric does not seem to be a natural vacuum solution outside of rigidly rotating, non-collapsed bodies (compare the remarks at the end of Sect. 2.2).

With the field equations completely solved and analyzed in order ω^2, we come to the ω^2 corrections of the extrinsic curvature tensors (according to (B.10)) of the mass shell, and to its energy-momentum tensor. According to

$$\frac{dr_S}{d\vartheta} = 2\omega^2 R^3 \, f \sin\vartheta \cos\vartheta,$$

the basis vector e_ϑ^μ obtains the correction $2\omega^2 R^3 \, f \sin\vartheta \cos\vartheta \, (0,1,0,0)$, and the normal vector n_μ the correction $-2\omega^2 R^3((X+1)^2/X^2) \, f \sin\vartheta \cos\vartheta \, (0,0,1,0)$. Of course, the basis vectors e_t^μ and e_φ^μ remain unchanged. For the components of the extrinsic curvature tensor we thus obtain

$$\overset{0}{K}_{ab} + (\omega R)^2 \overset{2}{K}_{ab} = n_\mu e_{a;b}^\mu$$

$$= \frac{(X+1)^2}{X^2} \left[e_{a,b}^r + {}^{(4)}\Gamma_{bc}^r e_a^c - 2\omega^2 R^3 \, f \sin\vartheta \cos\vartheta \, {}^{(4)}\Gamma_{bc}^\vartheta e_a^c \right].$$

$$\tag{B.33}$$

In principle, the components K_{ab} could now be given completely and explicitly. It turns out, however, that the (ϑ-independent) monopole terms get algebraically quite involved. Since the quadrupole terms (proportional to $\sin^2\vartheta$) are also by far the most interesting ones physically, we restrict ourselves to these parts, and denote them by ΔK_{ab}. In the first place, there appears a term $h'(X)$ in these expressions. However, this term is quite simply expressible by $f(X)$ and algebraic functions of X, due to a formula for $\overset{2}{U}_1(r,\vartheta)$ which results from a combination of the field equations (B.5) and (B.2):

$$h'(X) = -\frac{8(2X^2 - 1)}{3(X^2 - 1)^2} f - \frac{3\lambda^2(X - 1)}{8(X + 1)^7}.$$

$$\tag{B.34}$$

Herewith, the non-zero components of the tensor ΔK_{ab} take the forms

$$R\Delta K_{tt}^{+} = \frac{4X^3(X-1)}{(X+1)^5}\left[f + \lambda^2\frac{9X^2+14X+9}{64(X+1)^6}\right]\sin^2\vartheta,$$

$$\Delta K_{\vartheta\vartheta}^{+}/R = -\frac{X+1}{X^2(X-1)}\left[(5X^2-8X+5)f - \frac{9\lambda^2X(X-1)^2}{16(X+1)^6}\right]\sin^2\vartheta,$$

$$\frac{\Delta K_{\varphi\varphi}^{+}/R}{\sin^2\vartheta} = -\frac{X+1}{X^2(X-1)}\left[(3X^2-8X+3)f - \frac{9\lambda^2X(X-1)^2}{16(X+1)^6}\right]\sin^2\vartheta,$$

$$R\Delta K_{tt}^{-} = -\lambda^2\frac{X^4}{(X+1)^{10}}\sin^2\vartheta,$$

$$\Delta K_{\vartheta\vartheta}^{-}/R = -5\frac{(X+1)^2}{X^2}f\sin^2\vartheta,$$

$$\frac{\Delta K_{\varphi\varphi}^{-}/R}{\sin^2\vartheta} = -3\frac{(X+1)^2}{X^2}f\sin^2\vartheta. \tag{B.35}$$

With these results, to order ω^2, the quadrupole parts ΔS_b^a of the energy-momentum tensor take the form

$$8\pi R\Delta S_t^t = -\frac{8X^2}{(X+1)^3}\left[f + \frac{3\lambda^2X(3X^2-2X+3)}{64(X+1)^6(X-1)}\right]\sin^2\vartheta,$$

$$8\pi R\Delta S_\vartheta^\vartheta = \frac{2X^2}{(X+1)^3(X-1)}f\sin^2\vartheta,$$

$$8\pi R\Delta S_\varphi^\varphi = \frac{6X^2}{(X+1)^3(X-1)}\left[f + \frac{\lambda^2X^2}{4(X+1)^6}\right]\sin^2\vartheta. \tag{B.36}$$

All these quadrupole 'correction terms' to the static terms $\overset{0}{S}{}_a^a$ from (B.12) diverge in the collapse limit $X \to 1$, This is no surprise, since the stresses $\overset{0}{S}{}_\vartheta^\vartheta$ and $\overset{0}{S}{}_\varphi^\varphi$ already diverge there, and the whole expansion in powers of ω then breaks down. ΔS_t^t is negative for $X > 1.3$, i.e., in the whole physically realistic region. This makes the (equally negative) energy density $\overset{0}{S}{}_t^t$ bigger (in absolute value) at the equator than at the poles. $\Delta S_\vartheta^\vartheta$ is negative in the whole physical region $X > 1$, i.e., it makes the pressure $\overset{0}{S}{}_\vartheta^\vartheta$ from (B.12) smaller at the equator. ΔS_φ^φ is positive for

$X < 10$ and negative for $X > 10$. These results differ from the (presumably wrong) formulas (4.2)–(4.4) in Pfister and Braun (1985).

We also calculate the solution to order ω^3 because we would like to show explicitly that a mass shell with flat interior cannot rotate rigidly, and show explicitly what the differential rotation looks like. As already mentioned in the beginning of this appendix, to this order only the differential equation (B.6) for the function $\overset{2}{A}(r,\vartheta)$ is relevant, and its ω^3-contribution $\overset{2}{A}(r,\vartheta)$ obeys the equation

$$\overset{2}{A}_{11} + \frac{1}{r}\overset{2}{A}_1 + \frac{1}{r^2}\overset{2}{A}_{22} + \frac{3}{\overset{0}{W}}\left(\overset{0}{W}_1\overset{2}{A}_1 + \frac{1}{r^2}\overset{0}{W}_2\overset{2}{A}_2\right) - 4\overset{0}{U}_1\overset{2}{A}_1 = \overset{2}{V}_1\overset{0}{A}_1, \qquad \text{(B.37)}$$

with $\overset{2}{V} = 4\overset{2}{U} - 3\overset{2}{W}/\overset{0}{W}$. As also mentioned, the ϑ-dependence of $\overset{2}{A}(r,\vartheta)$ is only of the form $\sin^2\vartheta$. Inserting the functions $\overset{0}{W}(r,\vartheta)$ and $\overset{0}{U}(r)$ from (B.9) and (B.7), we see that the ansätze

$$\overset{2}{A}(r,\vartheta) = p(r) + q(r)(4 - 5\sin^2\vartheta), \quad \overset{2}{V}(r,\vartheta) = P(r) + Q(r)(4 - 5\sin^2\vartheta) \qquad \text{(B.38)}$$

leads to a separation of (B.37) of the form

$$\frac{d^2 p}{dy^2} - \frac{2}{y}\frac{dp}{dy} = \frac{dP}{dy}\frac{d\overset{0}{A}}{dy}, \qquad \text{(B.39)}$$

$$\frac{d^2 q}{dy^2} - \frac{2}{y}\frac{dq}{dy} - \frac{10q}{y^2(1-y)} = \frac{dQ}{dy}\frac{d\overset{0}{A}}{dy}, \qquad \text{(B.40)}$$

where, as for the terms of first order in ω, a change to the variable $y = 4x/(x+1)^2$, with $x = 2r/M$, is opportune. The homogeneous solutions of Eqs. (B.39) and (B.40) are found by standard methods of the theory of linear ordinary differential equations of second order (see, e.g., Morse and Feshbach 1953, Chap. 5.2), using the Wronskian determinant:

$$p_1 = 1; \qquad p_2 = y^3, \qquad \text{(B.41)}$$

$$q_1 = \frac{(1-y)(1-2y/3)}{y^2}, \quad q_2 = \bar{q}_2 + q_1\log(1-y), \text{ with}$$

$$\bar{q}_2 = \frac{1}{y} - \frac{7}{6} + \frac{y}{6} + \frac{y^2}{36} + \frac{y^3}{180}, \qquad \text{(B.42)}$$

with Wronskians $W_p = 3y^2$ and $W_q = y^2/90$, and where only the asymptotically decreasing solutions $p_2(y)$ and $q_2(y)$ are physically acceptable. The function $q_2(y)$ is positive in the whole physical region $0 < y \le 1$, but it is everywhere quite small:

From the maximal value $q_2(1) = 1/30$ it falls down to $q_2 = 0.01645$ already
at $y = 8/9$ ($x = 2$), and behaves asymptotically for $y \to 0$ like $y^5/630$.
The inhomogeneous solutions of (B.39) and (B.40) are again found according to
Morse and Feshbach (1953), Sect.5.2. Since the term $d\overset{0}{A}/dy = 3\lambda y^2/64$ essentially
cancels the Wronskians, we get

$$p_{\text{inh}} = \frac{3\lambda}{64} \int dy\, y^2 P(y);$$ (B.43)

$$q_{\text{inh}} = \frac{135\lambda}{32}\left[q_2 \int dy\frac{dQ(y)}{dy}q_1 - q_1 \int dy\frac{dQ(y)}{dy}q_2\right].$$ (B.44)

After inserting the functions $P(y)$ and $Q(y)$, according to (B.18), (B.21)
and (B.22), the integration constants in (B.43) and (B.44) have to be chosen such
that asymptotically decreasing solutions result. Since the ϑ-independent monopole
function $p(r)$ is algebraically quite involved and physically not so interesting, we
confine ourselves to the explicit presentation of the function $q(r)$ in its 'mixed'
dependence on x and y.

$$q = -\frac{\lambda}{40}\left\{96\beta_2 \frac{x(x-1)}{(x+1)^7} + \frac{\lambda^2}{2048X^2}y^5(1-y)\right.$$
$$\left. -3\gamma_2\left[y^2 - \frac{5}{6}y^3 + y(1-y)(1-\frac{y}{3})\log(1-y)\right] + \epsilon_2 q_2(y)\right\},$$ (B.45)

with a constant ϵ_2. This result coincides with Eq. (2.19) of Pfister and Braun (1986)
if one observes the different definition of the constants. It is remarkable that the
function $q(r)$ decreases asymptotically as r^{-5}, and therefore by two powers of r^{-1}
more than the function $\overset{0}{A}(r)$.

In order to guarantee the flatness of the interior solution also in order ω^3, the
exterior function $\overset{0}{A}(r)+(\omega R)^2\overset{2}{A}(r,\vartheta)$ has to join continuously to a constant interior
value $\overset{0}{A}_0+(\omega R)^2\overset{2}{A}_0$ at the shell position $r = r_S = R(1+(\omega R)^2 f \sin^2\vartheta)$. Inserting
$\overset{0}{A}(r)$ from Eq. (B.14) and $\overset{2}{A}(r,\vartheta)$ from Eq. (B.38) results in

$$\frac{q(X)}{\lambda} = -\frac{3X^3(X-1)}{5(X+1)^7}f(X).$$ (B.46)

The constant ϵ_2 is thus fixed at

$$\epsilon_2 = \frac{1}{q_2(Y)}\left\{3\gamma_2\left[Y^2 - \frac{5Y^3}{6} + Y(1-Y)(1-\frac{Y}{3})\log(1-Y)\right] - \frac{8(X-1)^2(2X-1)^2}{X^3(X+1)^2(3X-1)^2}\right\}.$$
(B.47)

Having now completely fixed the (quadrupole part of) function $\overset{2}{A}(r,\vartheta)$, we can calculate the order ω^3-quadrupole corrections to $K_{t\varphi}$ and S_t^φ, according to (B.10):

$$\Delta K_{t\varphi}^+ = \frac{(\omega R)^3 \lambda X}{(X+1)^6 (X-1)}$$

$$\left[(6X^3 - 20X^2 + 13X - 3)f - \frac{9\lambda^2 X(X-1)^2}{16(X+1)^5} - \frac{5(X+1)^8(X-1)}{2X^2}\frac{q'(X)}{\lambda}\right]\sin^4\vartheta;$$

$$\tag{B.48}$$

$$\Delta K_{t\varphi}^- = (\omega R)^3 \frac{X}{(X+1)^4} 3\lambda f \sin^4\vartheta. \tag{B.49}$$

$$8\pi R \Delta S_t^\varphi = \frac{\omega\lambda X^5}{(X+1)^{10}(X-1)}\left[(6X^3 - 26X^2 + 9X - 1)f\right.$$

$$\left.-\frac{3\lambda^2 X}{8(X+1)^5}(3X^2 - 2X + 3) - \frac{5(X+1)^8(X-1)q'(X)}{2\lambda X^2}\right]\sin^2\vartheta.$$

$$\tag{B.50}$$

By analogy with the analysis to first order of ω, with the determination of the constant λ in (B.17), we must now also demand to order ω^3 that the energy-momentum tensor S_ν^μ should represent a mass shell in purely axial rotation, i.e., that the eigenvector of $S_\nu^\mu u^\nu = -\rho u^\mu$ should have the form $u^\mu = u^0(1,0,0,\bar\omega)$. However, since for the quadrupole terms of this equation no free constant is now available (the constant ϵ_2 has already been determined by (B.47)), it turns out that this condition cannot be accomplished with a constant (ϑ-independent) angular velocity, but only with

$$\bar\omega = \omega[1 + (\omega R)^2 e(X)\sin^2\vartheta], \tag{B.51}$$

i.e., the mass shell with flat interior cannot rotate rigidly to order ω^3. By combining the above eigenvalue equations for $\mu = t$ and $\mu = \varphi$, thereby eliminating the mass density ρ, we get

$$S_t^t + \bar\omega S_\varphi^t = \frac{1}{\bar\omega}S_t^\varphi + S_\varphi^\varphi. \tag{B.52}$$

Inserting (B.51) and the expressions $\overset{1}{S}{}_\varphi^t$ (easily derivable from (B.16)), ΔS_t^t and ΔS_φ^φ from (B.36), and ΔS_t^φ from (B.50), we get the explicit (and non-zero!) expression

$$e(X) = \frac{X-1}{(X+1)^2(2X-1)(3X-1)}\left[-3(12X^3 - 21X^2 + 8X - 1)f - \right.$$

$$\left.-\frac{27(X+1)^5(X-1)^4(2X-1)}{X^5(3X-1)^2} + \frac{5(X+1)^8(2X-1)}{X^2}\frac{q'(X)}{\lambda}\right]. \tag{B.53}$$

(a) In the collapse limit $X \to 1$ the function $e(X)$ reaches the value zero. This result is in agreement with the work (de la Cruz and Israel 1968) where it was proven that spherical and rigidly rotating mass shells with flat interior can be the source of the Kerr metric only in the black hole limit $X \to 1$.

(b) In the region $X \in (1, 3.5)$ near the collapse limit the function $e(X)$ is negative. For $X > 3.5$ $e(X)$ is positive, i.e., the angular velocity of the equatorial parts of the mass shell with flat interior is somewhat higher than of the polar parts.

(c) In the weak field limit $X \to \infty$ the coefficient $e(X)$ reaches the non-zero and positive value $e(X \to \infty) = 227/27 \approx 8.4$.

These results differ considerably from the results in Fig.1 of Pfister and Braun (1986).

To conclude this appendix we would like to make some remarks concerning the metric for a rotating mass shell with flat interior to arbitrary order of the angular velocity parameter ω. The contribution to the metric function $W(x, \vartheta)$ outside the mass shell in order ω^{2n} is, according to the field equation $\Delta W = 0$, of the form

$$\overset{2n}{W} = \sum_{m=0}^{n} \overset{2n}{\beta}_m x^{-2m-1} \sin(2m + 1)\vartheta. \tag{B.54}$$

All coefficients $\overset{2n}{\beta}_m$ except $\overset{2n}{\beta}_0$ are fixed by the condition that this function continuously joins the interior function $W = e^{K_0} r \sin \vartheta$, as required by the flatness condition. The function $\overset{2n}{U}(x, \vartheta)$ is suitably written in the form

$$\overset{2n}{U} = \sum_{m=0}^{n} \overset{2n}{g}_m(x) P_{2m}(\cos \vartheta). \tag{B.55}$$

Then, due to the field equation (B.3), the functions $\overset{2n}{g}_m(x)$ have to satisfy the differential equations

$$\frac{d^2 \overset{2n}{g}_m}{dx^2} + \frac{2x}{x^2 - X^2} \frac{d \overset{2n}{g}_m}{dx} - \frac{2m(2m + 1)}{x^2} \overset{2n}{g}_m = \overset{2n}{I}_m, \tag{B.56}$$

where $\overset{2n}{I}_m$ are inhomogeneities, given by the lower order solutions. Since each of Eqs. (B.56) has only one asymptotically decreasing homogeneous solution, the function $\overset{2n}{U}(x, \vartheta)$ outside the mass shell contains $n + 1$ integration constants. If we demand that $\overset{2n}{U}$ decreases faster than x^{-1} (no correction of order ω^{2n} to the total mass!), only n integration constants remain. These n constants are, however, completely determined by the condition of continuity between $\overset{2n}{U}(x, \vartheta)$ and some interior constant U_0. After inserting $\overset{2n}{W}$ into the field equations (B.4) and (B.5),

$\overset{2n}{K}(x, \vartheta)$ has to satisfy two first-order linear differential equations which completely determine the (asymptotically decreasing!) function $\overset{2n}{K}$. Representing $\overset{2n}{K}$ in the form

$$\overset{2n}{K}(x, \vartheta) = \sum_{m=0}^{n} \overset{2n}{k}_m(x) \sin^{2m} \vartheta,$$

the $n+1$ continuity conditions for $\overset{2n}{k}_m(x)$ at the shell position are now just realizable with the help of the remaining free constants, viz., the constants $\overset{2n}{\beta}_0$ from $\overset{2n}{W}$, and n constants $\overset{2n}{f}_m$ from the order $2n$ correction to the shell geometry

$$r_S = R\left(1 + (\omega R)^2 f \sin^2 \vartheta + \cdots + (\omega R)^{2n} \sum_{m=1}^{n} \overset{2n}{f}_m \sin^{2m} \vartheta\right). \tag{B.57}$$

In total, we see that to any even order ω^{2n} there exists (for given M, R, and ω) exactly one rotating mass shell with flat interior, starting from a spherical shell to order ω^0.

In uneven orders of ω, (B.6) for $\overset{2n}{A}(y, \vartheta)$ has to be considered outside the mass shell. In an extension of (B.38), it is advantageous to represent $\overset{2n}{A}$ by the Jacobi polynomials $P_{2m}^{(1,1)}(\cos \vartheta)$ (see, e.g., Gradstein and Ryzhik 1981, p. 443).

$$\overset{2n}{A}(y, \vartheta) = \sum_{m=0}^{n} \overset{2n}{p}_m(y) P_{2m}^{(1,1)}(\cos \vartheta). \tag{B.58}$$

Then the functions $\overset{2n}{p}_m(y)$ have to fulfil the differential equations

$$\frac{d^2 \overset{2n}{p}_m}{dy^2} - \frac{2}{y} \frac{d \overset{2n}{p}_m}{dy} - \frac{2m(2m+3)}{y^2(1-y)} \overset{2n}{p}_m = \overset{2n}{P}_m, \tag{B.59}$$

with given inhomogeneities $\overset{2n}{P}_m$. Again, each of Eqs. (B.59) has only one asymptotically decreasing homogeneous solution, so that $\overset{2n}{A}$ contains $n + 1$ integration constants. From these $n + 1$ numbers, n are fixed by the condition of continuity of $\overset{2n}{A}$ with some interior constant $\overset{2n}{A}_0$. With $\overset{2n}{A}$ known, the energy-momentum tensor components S_φ^t and S_t^φ are given up to order $2n + 1$, and they have to satisfy the condition (B.52) of axial rotation, which, to order $2n + 1$ represents $n + 1$ equations for the terms $\sin^{2m} \vartheta$ with $m = 0, 1, \ldots, n$. For these $n + 1$ equations, we have the integration constant from $\overset{2n}{p}_0(y)$, and n constants $\overset{2n}{e}_m$ from the order $2n$ correction to the angular velocity

$$\bar{\omega} = \omega \left(1 + (\omega R)^2 e \sin^2 \vartheta + \cdots + (\omega R)^{2n} \sum_{m=1}^{n} \overset{2n}{e}_m \sin^{2m} \vartheta \right). \tag{B.60}$$

This shows that also in uneven orders of ω there is exactly one solution for a rotating mass shell with flat interior. However, the analytic form of the solutions of orders $n > 3$ is surely much more complicated than the above (already involved) solutions of orders $n = 2$ and $n = 3$, because, e.g., the differential equations for these solutions already contain factors $\log((x - 1)/(x + 1))$ or $\log(1 - y)$.

References

de la Cruz, V., Israel, W.: Spinning shell as a source of the Kerr metric. Phys. Rev. **170**, 1187–1192 (1968)

Gradstein, I.S., Ryzhik, I.M.: Tables of Series, Products, and Integrals, vol. 2. Harri Deutsch, Thun (1981)

Hartle, J.B.: Slowly rotating relativistic stars I. Equation of structure. Astrophys. J. **150**, 1005–1029 (1967)

Israel, W.: Singular hypersurfaces and thin shells in general relativity. Nuovo Cimento **44**, 1–14 (1966)

Morse, P.M., Feshbach, H.: Methods of Theoretical Physics, vol. I. McGraw-Hill, New York (1953)

Pfister, H., Braun, K.H.: Induction of correct centrifugal force in a rotating mass shell. Class. Quant. Grav. **2**, 909–918 (1985)

Pfister, H., Braun, K.H.: A mass shell with flat interior cannot rotate rigidly. Class. Quant. Grav. **3**, 335–345 (1986)

Pfister, H., Frauendiener, J., Hengge, S.: A model for linear dragging. Class. Quant. Grav. **22**, 4743–4761 (2005)

Thirring, H.: Über die Wirkung rotierender ferner Massen in der Einsteinschen Gravitationstheorie. Phys. Zs. **19**, 33–39 (1918)

Name Index

© Springer International Publishing Switzerland 2015
H. Pfister, M. King, *Inertia and Gravitation*, Lecture Notes in Physics 897,
DOI 10.1007/978-3-319-15036-9

Subject Index

Absolute space, viii, 5–7, 9, 11, 25, 29, 120
Absolute time, viii, 5–7, 9, 11, 23, 25, 29, 120
Action at a distance, 39, 120
Action-at-a-distance force, 3, 4
Action principle, 85
ADM energy, 96, 97
ADM mass, 106, 107
Affine connection, 19, 21–24, 34–36, 49, 50, 71
Affine manifold, 21, 28, 29
Affine parameter, 71, 73
Affine space, 22
Affine structure, 19, 21–23, 28, 29, 34, 36, 55, 69–75
Anti-deSitter solution, 98
Antidragging effect, 132
Antiparticle, 108, 109

Banach fixed point theorem, 42, 94, 108
Bianchi identities, 75, 85, 90
  contracted, 75, 91, 92
Big Bang singularity, 99, 110, 111
Blackbody radiation, 109
Black hole, vi, 51, 86, 88, 99–102, 105, 106, 111, 144
  geometric inequalities, 106
  hoop conjecture, 106, 107
  horizon, 98, 103
  horizon area, 104, 105
  information loss, 105
  no-hair theorem, 103, 104
  Penrose inequality, 106

physics, 101
uniqueness theorem, 103
Body Alpha, 6
Bondi energy, 97
Boundary value problem, 94

Cauchy problem, vi, 92–94
Cauchy surface, 98
Causal structure, 24
Centrifugal effect, 104
Centrifugal field, 84, 122
Centrifugal force, 2, 120, 121, 123–125, 127, 128
Chandrasekhar mass, 100, 108
Chern–Pontryagin invariant, 143, 144
Christoffel symbols, 50, 68, 73, 75, 86, 90
Collapse limit, 127, 130, 131, 133
Compact star, 86
Conformal connection, 59, 71
Conformal curvature, 81
Conformal geometry, 50
Conformal structure, 50, 52–64, 66, 67, 69, 73–75
Constraints, 92–94, 134, 135, 137
Coriolis field, 122
Coriolis force, 120–125, 127, 128
Cosmic censorship hypothesis, 99, 106
Coulomb's law, 4, 40, 91, 120, 143
Covariant derivative, 59, 66
Critical exponent, 107
Critical phenomena, 107
Curvature, 37
Curvature tensor, 73
  Riemann, 75

© Springer International Publishing Switzerland 2015
H. Pfister, M. King, *Inertia and Gravitation*, Lecture Notes in Physics 897,
DOI 10.1007/978-3-319-15036-9